Customized Version of

MATHEMATICS BEYOND THE NUMBERS

by George T. Gilbert
and Rhonda L. Hatcher
Designed Specifically
for Valencia College

Kendall Hunt
publishing company

Kendall Hunt
publishing company

www.kendallhunt.com
Send all inquiries to:
4050 Westmark Drive
Dubuque, IA 52004–1840

ISBN 978-1-4652-0417-2

Printed in the United States of America
10 9 8 7 6 5 4 3 2 1

CONTENTS

MATHEMATICS BEYOND THE NUMBERS

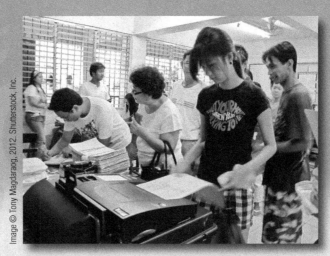

Voters in the election for President of the Philippines.

Voting methods lie at the intersection of political science, economics, and mathematics. Whether a group is electing a leader or choosing a course of action, its decision may depend on the voting method that is used as well as on the issues and the preferences of the voters. In this chapter we focus our attention on voting to reach a single decision—for example, electing an officer. We do not consider multiple, linked decisions.

With the two-party system dominating elections in the United States, voters often face the simplest situation: choose between two candidates, and the one receiving the most votes wins. The procedure of counting ballots treats voters equally and choices equally. Barring a tie, the outcome of an election with two alternatives is always clear and unequivocal. This is the larger purpose of democratic government—to provide a means of reaching public decisions that unites the people behind a reasonable course of action.

However, many elections require making a decision from among more than two choices. For example, the 1968 presidential election featured a strong challenge from third-party candidate George Wallace. In this election, the Democratic nominee, Hubert Humphrey, received 42.7% of the vote, whereas the Republican nominee and winner of the election, Richard Nixon, received 43.4%. If George Wallace, who received 13.5% of the vote, had not run, the election might have yielded a different winner. Presidential primaries and local elections often require voters to choose among a long list of candidates. If there are only two alternatives to choose between, we know how to proceed: majority rules. The procedures for arriving at a group decision when the group has at least three options to choose from are not so simple and raise interesting questions. In this chapter we look at some of these procedures.

Lining up to cast ballots in Poland.

VOTING METHODS

1.1   PLURALITY AND RUNOFF METHODS

The simplest, and perhaps most common, method of deciding among three or more alternatives is the plurality method, described as follows.

The Plurality Method

With the **plurality method**, each voter selects one candidate or choice on the ballot. The winner is the candidate or choice with the most votes.

The winner in an election decided by the plurality method is said to have won by a plurality or, in Great Britain, a relative majority. Note that in an election with three or more candidates, a candidate need not have a majority to win by a plurality. The plurality method is illustrated in the following example.

EXAMPLE 1   *Plurality Method*

A committee must vote for a main dish to be served at a banquet. The choices are chicken, pork, fish, or beef. The committee decides to use the plurality method, and their ballots read as follows: chicken, pork, beef, chicken, fish, beef, fish, beef, chicken, chicken, beef, pork, beef.

(a)   Which main dish did they choose?
(b)   Just before counting the vote, the committee realizes that pork is not one of the options offered by the caterer. If those who voted for pork replace their ballots, which main dishes could be the group's selection?

SOLUTION:

(a)   The person assigned to count the votes would probably sort the ballots into four piles and find that the tally of the original ballots is beef 5, chicken 4, pork 2, fish 2, so beef is chosen as the main dish.
(b)   There are only 2 votes for pork, so fish could get at most 4 votes, not catching up to beef. On the other hand, 1 additional vote for beef or 2 additional votes for chicken would result in that dish being selected. Thus, beef or chicken could be selected.   ∎

If one of the people voting for pork or fish in Example 1 voted for chicken instead, what would have been the outcome of the election? The result would have been a tie between beef and chicken with 5 votes each. With a large number of voters, ties are uncommon. However, when the number of voters is small, ties can easily occur under all of the voting methods we will study.

EXAMPLE 2 *Minimum Votes Needed to Win*

Suppose that 60 votes are cast in an election among four candidates: Hughes, Dunbar, Torres, and Ishmael. After the first 40 votes are counted, the tallies are as follows:

Hughes	8
Dunbar	17
Torres	12
Ishmael	3

What is the minimal number of the remaining votes Dunbar can receive and be assured of a win?

SOLUTION:

There are 20 votes left. To win the election, Dunbar must have the most votes after the last 20 votes are cast. Torres is in second place behind Dunbar after the first 40 votes are cast. In the worst-case scenario for Dunbar, Torres would get all the votes that did not go to Dunbar. Torres needs 5 votes to catch up with Dunbar, and then 8 of the remaining 15 votes to beat Dunbar. Thus, Torres would need a total of 5 + 8 = 13 votes. We can conclude that if Dunbar gets 8 of the last 20 votes, then neither Torres, Hughes, nor Ishmael can catch up, and Dunbar would be assured of a win.

Alternatively, we can solve this problem algebraically. Let x be the number of votes Dunbar needs to ensure at least a tie for first place. If Torres gets all the votes that do not go to Dunbar and the race ends with a tie between Torres and Dunbar, we must have

$$17 + x = 12 + (20 - x).$$

Solving this equation for x, we get

$$17 + x = 32 - x,$$
$$x + x = 32 - 17,$$
$$2x = 15,$$
$$x = \frac{15}{2} = 7.5.$$

We conclude that 7 more votes are not enough to assure a win, but that 8 votes will do the job, ensuring that Dunbar has more votes than any other candidate. ∎

If the plurality method is used in an election involving more than two candidates and one of the candidates receives more than 50% of the votes, then that candidate clearly wins the election and it can be reasonably argued that this winner is the popular

branching OUT

THE ELECTORAL COLLEGE

Americans elect their presidents, not by popular vote, but instead through the Electoral College. Under this system, each state has a number of electoral votes equal to the number of Representatives from the state plus two for the state's two Senators. In addition, the District of Columbia is entitled to three electoral votes. The votes are cast by individuals called electors. If a candidate wins a *majority*, not merely a plurality, of the electoral votes, then the candidate is declared the winner. If no candidate wins, then the election is handed over to the House of Representatives, who vote for one of the top three contenders.

The method of electing the president through the Electoral College was adopted at the Constitutional Convention of 1787 after much debate. The plan left the method by which a state's electors would be chosen to the discretion of the state legislatures. Initially, states used a variety

of methods for selecting electors, but gradually the states moved toward a standard system of choosing the presidential electors by a statewide, winner-take-all popular vote. Under this system, whichever candidate wins a plurality of the popular vote in the statewide election is entitled to *all* of the state's electoral votes. The only current exceptions are Maine and Nebraska, which allocate electors corresponding to Representatives on a district-by-district basis. However, because electors merely pledge to support a particular candidate, they occasionally break the pledge and vote for another candidate. As recently as 2004, an anonymous Minnesota elector pledged to Democrat John Kerry instead voted for Kerry's running mate, John Edwards, for president rather than Kerry.

Because the electoral votes of a state normally go to the one candidate who wins a plurality of the popular vote in the state, no matter how close the

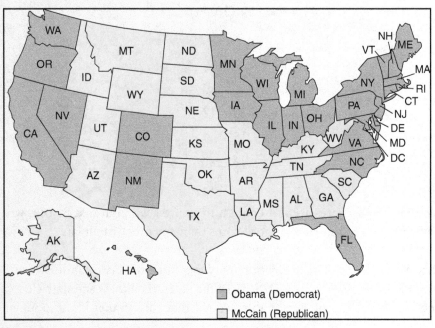

- ☐ Obama (Democrat)
- ☐ McCain (Republican)

The states won by Obama and McCain in the 2008 election.

Image © Dr_Flash, 2012. Shutterstock, Inc.

tally, the electoral vote totals often exaggerate the winning margin. For instance, in the 2008 election, Barack Obama beat John McCain by 52.9% to 45.5% in the national popular vote, but in the electoral vote the split was 365 to 173 with Obama taking 67.8% of the electoral votes.

Third-party candidates can have significant support across the country, but will only win electoral votes in states in which they win a plurality of the popular vote. For instance in the 1992 election, even though Ross Perot captured 19% of the popular vote, he did not win any electoral votes because he did not have a plurality of the votes in any state.

It is possible for a candidate who has not even won a plurality of the national popular vote to

win the election in the Electoral College. This has happened four times in the history of the country, with the elections of John Quincy Adams in 1824, Rutherford B. Hayes in 1876, Benjamin Harrison in 1888, and George W. Bush in 2000. (For more on the controversial election of Hayes, read Branching Out 2.1.) John Quincy Adams was elected president even though he failed to win a plurality of the Electoral College votes. He had only 84 electoral votes to the 99 for Andrew Jackson. But this election included two other strong candidates: Henry Clay, who captured 37 electoral votes, and William H. Crawford, with 41 votes. Because no candidate had a majority of the 261 electoral votes, the House of Representatives decided the election and gave the victory to John Quincy Adams.

choice. However, it is often the case in plurality elections that no candidate receives more than 50% of the votes. For instance, in the 2006 Texas gubernatorial election two independent candidates garnered significant support and Rick Perry was elected with a 39% plurality. Given the importance of popular support, the question arises whether plurality is the best voting method to use when an election has more than two candidates.

A central problem with the plurality method is illustrated by the following example. Suppose that an election between two very different candidates, Smith and O'Neil, would result in Smith comfortably defeating O'Neil 60% to 40%. Now suppose that two candidates very similar to Smith, whom we will call Smyth and Smythe, are also in the election and that there is not much to distinguish Smith, Smyth, and Smythe. In this case, 60% of the voters would prefer either Smith, Smyth, or Smythe to O'Neil. Thus, in an election with all four candidates, O'Neil would probably win with 40% of the vote, whereas Smith, Smyth, and Smythe individually would run far behind with roughly 20% each. With many candidates, the plurality system makes it fairly likely that an extreme candidate with a small but intense base of support will defeat a hoard of candidates with broad support and similar views. The propensity of the Democratic Party to select a liberal candidate and of the Republican Party to select a conservative candidate through their presidential primaries and caucuses, rather than more moderate candidates, supports this observation. Among the more memorable examples of this are the Republican Party's nomination of Barry Goldwater in 1964 and the Democratic Party's nomination of George McGovern in 1972.

A second problem with the plurality method is the dilemma that a supporter of a "weak" candidate faces when there are more than two candidates running. Should the

2008 Democratic Party Convention.

voter cast a ballot for his or her first choice, a "wasted" vote, or switch to one of the serious contenders in an effort to affect the outcome of the election? The incentive to switch is an incentive to vote strategically. When a person votes in a way that does not reflect his or her true preferences in an attempt to improve the outcome of the election from that person's point of view, we call this **strategic voting** and say that the voter is voting strategically.

Runoff Elections

Because of shortcomings in the plurality method, many elections, particularly local elections in which races often have several candidates, have a provision for a runoff election as described next.

Runoff Election

In a **runoff election**, a plurality vote is taken first. If one of the candidates has more than 50% of the votes, then that candidate wins. If none of the candidates receives a majority of the votes, then a second plurality election is held with a designated number of the top candidates. The process is repeated until one candidate has more than 50% of the votes.

In most cases, the top two candidates from the first vote run in the runoff election. Having a provision for a runoff helps in elections in which an extreme candidate wins the plurality vote with fewer than 50% of the votes. A runoff election also lessens the incentive to switch one's vote from a preferred candidate to a more electable one, because there will be a second chance to vote unless one candidate receives a majority of all votes, in which case such a decision is irrelevant.

EXAMPLE 3 *1992 Presidential Election*

George Bush, Ross Perot, and Bill Clinton debating during the 1992 presidential election campaign.

In the 1992 U.S. presidential election, Bill Clinton received 44,908,254 votes, George Bush received 39,102,343, and Ross Perot received 19,741,065, with some other votes going to other minor candidates. (Recall, however, that U.S. presidents are actually elected through the Electoral College.) Suppose there had been a runoff election between Clinton and Bush. Assuming the same people voted in the original election and the runoff and ignoring the voters supporting minor candidates, answer the following:

(a) In a runoff election between Clinton and Bush, what percentage of Perot supporters would need to vote for Clinton in order for Clinton to receive a majority of the popular vote?

(b) In a runoff election between Clinton and Bush, what percentage of Perot supporters would need to vote for Bush in order for Bush to receive a majority of the popular vote?

(c) Polls showed that Perot took support from Clinton and Bush in roughly equal numbers. How likely does it seem that George Bush could have won the popular vote without Ross Perot in the election?

SOLUTION:

(a) The total number of votes cast for each candidate was as follows:

Clinton	44,908,254
Bush	39,102,343
Perot	<u>19,741,065</u>
Total	103,751,662

Because the total is an even number, a majority is 1 more than half of the total or

$$\frac{1}{2}(103,751,662) + 1 = 51,875,831 + 1 = 51,875,832 \text{ votes.}$$

We assume that all of Clinton's and Bush's supporters in the original election will continue to support them in the runoff. Because Clinton received 44,908,254 votes in the original election, we see that he would need

$$51,875,832 - 44,908,254 = 6,967,578$$

of the Perot supporters' votes to have a majority. This is

$$\frac{6,967,578}{19,741,065} \approx 0.3529 = 35.29\%$$

of the Perot supporters.

(b) We could do a calculation similar to the one we did in part (a), but it is easier to observe that if Clinton needs more than 35.29% of the Perot supporters to receive a majority of the popular vote in the runoff, then Bush would need more than 100% − 35.29% = 64.71% to win in the runoff.

(c) In part (b), we saw that more than 64.71% of Perot supporters must switch to Bush in order for Bush to win in the runoff election. Because only about 50% of Perot's supporters would have switched to Bush if Perot had dropped out of the election, we can conclude that it does not seem likely that Bush would have won the popular vote had Perot dropped out.   ■

Preference Rankings

In most voting situations, each voter has an order of preference of the candidates. Such an ordering is called a **preference ranking**. The voter may have to think hard to arrive at this preference ranking, and it may change quickly and easily with time and circumstance. In some voting methods, such as Borda's method which we will see in the next section, the voters actually write their preference rankings on their ballots. In plurality elections, this is not the case. However, if we know the preference rankings of the voters, we can analyze how an election would come out under various voting methods.

It is possible for a voter to be indifferent between two or more alternatives—for instance, if the voter has no knowledge whatsoever of several of the candidates. We assume that whenever indifference among choices exists, the voter will arbitrarily, perhaps randomly, order the choices. Thus, we assume each voter has a preference ranking that orders all candidates from most preferred to least preferred.

We record the preference rankings of all the voters in an election by using a table such as Table 1.1.

TABLE 1.1 Example of Preference Rankings

	Number of Voters			
	6	3	1	4
Apples	1	2	2	3
Oranges	3	1	3	2
Pears	2	3	1	1

branching OUT

CHOOSING THE SITE OF THE 2012 SUMMER OLYMPICS

The Olympics have always been political, and politics plays a role in the selection of sites for the Olympic games. This is illustrated by the vote for the site of the 2012 Summer Olympics. The selection was made in 2005 by 104 members of the International Olympic Committee. The cities in contention were London, Madrid, Moscow, New York, and Paris. Because of past abuses, there were numerous restrictions on lobbying of voting members. After the initial evaluation reports, Paris was viewed as the frontrunner, followed closely by London.

The voting process calls for a plurality election, with a runoff between all of the candidates except the one in last place. This voting method is known as the *Hare method*. The process continues until a city has a majority of the votes. In the first round of voting the tallies were as follows:

First Round	
London	22
Paris	21
Madrid	20
New York	19
Moscow	15

Thus, Moscow was eliminated from the competition. The second round of votes yielded the following:

Second Round	
Madrid	32
London	27
Paris	25
New York	16

Source: International Olympic Committee.

There were 3 more voters in this round than in the first round. We see that, most likely, most of those voting for Moscow in the first round switched to Madrid. Interestingly, New York lost 3 votes from the first round. Such a thing has also happened in past elections. With New York eliminated, the third round votes were:

Third Round	
London	39
Paris	33
Madrid	31

Note the leapfrogging between London and Madrid. The final round of voting was:

Fourth and Final Round	
London	54
Paris	50

Note that Madrid lost one vote between the second and third rounds. Though not unusual, in this case it was later alleged by a senior International Olympic Committee member, Alex Gilady, that one voter had mistakenly voted for Paris over Madrid. He went on to speculate that Madrid would have defeated Paris in a tiebreaker round to determine the final round opponent for London and that Madrid would have then defeated London to become the 2012 host site.

Image © Pete Niessen, 2012. Shutterstock, Inc.

The three "candidates" in this election are apples, oranges, and pears, and they are listed in the first column on the left. In the next column we see a possible ranking, where apples are ranked first, pears second, and oranges third, as indicated by the numbers recorded in the rows corresponding to each of the fruits. The boldface number at the top of this column indicates the number of voters who have this particular preference ranking. In this case, there were six such voters. The remaining columns include other possible preference rankings and the number of voters with each ranking. We see by adding across the first row that there are a total of 14 voters. Notice that, in this example, not every possible ranking is included. For instance, no voters had a preference ranking of apples first, oranges second, and pears third, or a ranking of oranges first, pears second, and apples third.

Preference rankings have been used sporadically in the United States going back to 1912 party primaries in Florida, Indiana, Maryland, Minnesota, and Wisconsin. They are currently used in local elections in San Francisco; Basalt, Colorado; Ferndale, Michigan; Berkeley, California; Minneapolis; Takoma Park, Maryland; Oakland, California; Santa Fe, New Mexico; Memphis, Tennessee; Telluride, Colorado; Saint Paul, Minnesota; San Leandro, California; Portland, Maine; and in certain state-wide judicial and municipal elections in North Carolina.

Throughout this chapter we make some assumptions about preference rankings. Our first assumption is that if a voter has ranked one candidate higher than another, then if the voter must choose between those two candidates, the voter would choose the higher-ranked one. We also assume that the order of the preferences is not changed by the elimination of one or more candidates—for example, in the case of a runoff election.

Let's now look at an example of how the results of an election decided by a plurality with a runoff can be worked out if we know the preference rankings of the voters.

EXAMPLE 4 *Election Results from Preference Rankings*

Eleven members of a department are voting on which day to schedule their weekly meetings. Their choices are Tuesday, Wednesday, Thursday, and Friday, and their preference rankings are listed in Table 1.2.

TABLE 1.2 Preference Rankings for Meeting Day

| | *Number of Voters* | | | | | | |
	1	**3**	**1**	**1**	**1**	**2**	**2**
Tuesday	1	1	4	3	2	3	4
Wednesday	2	3	1	1	4	2	2
Thursday	4	4	2	4	1	1	3
Friday	3	2	3	2	3	4	1

(a)   In a plurality election, which day would be selected?

(b)   In a plurality election with a runoff between the top two finishers, which day would be selected?

(c)   In a plurality election with a runoff between the top two finishers, could the two voters who ranked Friday first achieve a preferable outcome by voting strategically if the others voted as indicated in the table?

SOLUTION:

(a)   To determine the winner of a plurality election, we need only count the first-place rankings. The preference rankings with Tuesday ranked first are shaded in blue in Table 1.3.

TABLE 1.3 Those Voting for Tuesday in a Plurality Election

			Number of Voters				
	1	3	1	1	1	2	2
Tuesday	1	1	4	3	2	3	4
Wednesday	2	3	1	1	4	2	2
Thursday	4	4	2	4	1	1	3
Friday	3	2	3	2	3	4	1

We see that a total of 4 voters ranked Tuesday first and would therefore vote for Tuesday in a plurality election. Similarly, Wednesday would get 2 votes, Thursday 3 votes, and Friday 2 votes. Therefore, Tuesday would win.

(b)   The runoff election choices would be Tuesday and Thursday, with each voter choosing whichever of these two days he or she ranks higher. In Table 1.4, the preference rankings in which Tuesday is ranked above Thursday are shaded blue. These 5 voters would vote for Tuesday in the runoff. The preference rankings in which Thursday is ranked above Tuesday are shaded pink, and these 6 voters would vote for Thursday. Therefore, Thursday would win with 6 votes, with Tuesday receiving 5 votes in the runoff.

TABLE 1.4 Runoff Election Between Tuesday and Thursday

			Number of Voters				
	1	3	1	1	1	2	2
Tuesday	1	1	4	3	2	3	4
Wednesday	2	3	1	1	4	2	2
Thursday	4	4	2	4	1	1	3
Friday	3	2	3	2	3	4	1

(c)   Recall that to vote strategically means to vote in a way that does not reflect one's true preferences, attempting to obtain an outcome preferable to that obtained when voting according to one's true preferences. These two voters prefer both Friday and Wednesday to Thursday, so they prefer that either of these two days wins instead of Thursday. There is no point in voting strategically in the runoff. However, if they vote for Wednesday instead of voting for Friday on the first ballot, then the results of the first ballot would be Tuesday 4, Wednesday 4, Thursday 3, and Friday 0. Therefore, the runoff would be between Tuesday and Wednesday. In the runoff, if all of the voters vote as expected from their preference rankings, then Tuesday would get 5 votes, and Wednesday would get 6 votes and win the election. Thus, we see that our two strategic voters can achieve the preferable outcome of meeting on Wednesday. ∎

Sometimes the breakdown of voters according to their preference rankings is given in terms of a percentage of voters rather than a number of voters. We look at an example of this kind next.

EXAMPLE 5 *Preference Rankings Broken Down by Percentage*

Suppose that a university is considering whether to expand, decrease, or maintain the current level of career counseling service. Student fees are affected by the level of service, with more service resulting in higher fees. The preference rankings of the student body regarding the level of service are as follows:

	Percentage of Voters			
	37	23	21	19
Expand	1	3	2	3
Decrease	3	1	3	2
Maintain	2	2	1	1

(a)   Which option would the students choose using the plurality method?

(b)   Which option would they choose using the plurality method followed by a runoff between the first- and second-place finishers?

SOLUTION:

(a)   Counting only the first-place votes, we see that in a plurality election 37% would vote to expand service, 23% would vote to decrease service, and 21% + 19% = 40% would vote to maintain service at its current level. Therefore, maintaining the level of service would win.

(b)   The runoff would be between maintaining service at its current level and expanding service. Because all those with decreasing service as a first choice had maintaining the level of service as a second choice, the results of the runoff election would be 37% for expanding services and 40% + 23% = 63% for maintaining service at its current level. Once again, we see that maintaining the level of service would win. ■

We saw that when there are more than two candidates, the plurality method can result in an outcome that is less desirable to a majority of the voters than some other particular outcome. We will see later that the same problem can occur with the plurality method with a runoff. In addition, both methods are susceptible to manipulation by strategic voting. Neither of these methods allows voters to express their relative preferences among *all* the candidates. In the next three sections we look at voting methods that allow voters to do this.

E X E R C I S E S / FOR SECTION 1.1

1. A group of friends is deciding what type of food to go out for on Sunday night. The votes are Mexican, Italian, Chinese, Mexican, Italian, French, Chinese, Chinese, Indian. Which type of food will they eat if the decision is by plurality?

2. A committee decides to elect a chair by a plurality election. The three candidates are Gorman, Hwang, and Page. The ballots read Hwang, Hwang, Page, Gorman, Page, Page, Hwang, Gorman, Hwang, Page, and Hwang. Who wins the election?

3. Suppose there are 100 votes cast in an election among three candidates: Flores, Payne, and Bronowski. The election is to be decided by plurality. After the first 70 votes are counted, the tallies are as follows:

Flores	31
Payne	23
Bronowski	16

    (a) What is the minimal number of remaining votes Flores needs to be assured of a win?
    (b) What is the minimal number of remaining votes Payne needs to be assured of a win?
    (c) What is the minimal number of remaining votes Bronowski needs to be assured of a win?

4. Suppose there are 80 votes cast in an election among three candidates—Donahue, Garza, and Weis—to be decided by plurality. After the first 55 votes are counted, the tallies are as follows:

Donahue	24
Garza	18
Weis	13

    (a) What is the minimal number of remaining votes Donahue needs to be assured of a win?

(b)   What is the minimal number of remaining votes Garza needs to be assured of a win?

(c)   What is the minimal number of remaining votes Weis needs to be assured of a win?

5. If 302 votes are cast, what is the smallest number of votes a winning candidate can have in a five-candidate race that is to be decided by plurality?

6. If 203 votes are cast, what is the smallest number of votes a winning candidate can have in a four-candidate race that is to be decided by plurality?

7. The 1824 presidential election included four competitive candidates, all from the Democratic-Republican Party. Andrew Jackson received 151,271 votes, John Quincy Adams received 113,122 votes, Henry Clay received 47,531 votes, William H. Crawford received 40,856 votes, and 13,053 voters voted for other candidates. Because Jackson failed to get a majority in the Electoral College, the House of Representatives decided the election by choosing John Quincy Adams.

(a)   If those voting for Crawford and miscellaneous candidates had voted for one of the top three, who could have won a plurality of the votes?

(b)   What percentage of these voters would Jackson need to ensure a plurality?

8. Suppose that three candidates—Rosen, Brown, and Wheatley—are running in an election that will be decided by the plurality method with a runoff between the top two finishers if none of the candidates receives a majority of the votes. The results of the first ballot are given next.

Rosen	2,346
Brown	5,784
Wheatley	6,230

In a runoff election between Wheatley and Brown, what percentage of Rosen supporters would need to vote for Brown in order for Brown to win the election?

9. In September and October 2011, members of Alberta's Progressive Conservative party voted to decide the Canadian province's next Premier. The election was to be decided by the plurality method with a runoff among the top three candidates if none of the candidates received a majority of the votes on the first ballot. On the second ballot, voters also picked a second choice so that there could be an instantaneous runoff if none of the three remaining candidates received a majority of the votes. The results of the first ballot held on September 17 are given next.

Gary Mar	24,195
Alison Redford	11,129
Doug Horner	8,635
Ted Morton	6,962
Rick Orman	6,005
Doug Griffiths	2,435

On October 1, a runoff election between Mar, Redford, and Horner was held, and the results were as follows.

Gary Mar	33,233
Alison Redford	28,993
Doug Horner	15,950

What percentage of Horner supporters would have needed to vote for Mar in order for Mar to win the final runoff? (As it turns out, Redford became Alberta's first female Premier with 37,104 votes to Mar's 35,491 votes.)

10. The procedure for enacting legislation in the U.S. House of Representatives and Senate is to consider amendments to a bill one at a time, and then to vote on passage of the entire bill at the end. In 1955, the Senate considered a highway appropriations bill introduced by Albert Gore of Tennessee, father of the future Vice President and Nobel Peace Prize winner. One amendment was to remove the so-called Davis–Bacon clause that mandated fair pay standards for workers on federal construction projects. The Senate could be divided into three groups, none forming a majority. One group, primarily Republicans, favored no bill of any sort, but favored keeping the Davis–Bacon clause in the bill. A second group, mainly Southern Democrats, favored the bill if and only if the Davis–Bacon clause were removed. The third group, mainly other Democrats, favored the bill in either form, but preferred inclusion of the Davis–Bacon clause.

    (a) What would have resulted from a vote on the amendment and then on the bill?

    (b) Could one of the three groups obtain a result preferable to the outcome in (a) by voting strategically on one of the votes?

    (What actually happened is that the Democratic leadership of the Senate, which included future President Lyndon Johnson, was able to remove the clause without a formal vote, and thus secured passage of the highway bill.)

11. Nine members of a committee must decide what kind of flooring to install in a community room. The preference rankings of the committee members are listed here.

	Number of Voters				
	3	1	2	1	2
Carpet	1	1	3	2	3
Ceramic tile	2	3	1	3	2
Wood	3	2	2	1	1

    (a) Which type of flooring would be chosen using the plurality method?

    (b) Which type of flooring would be chosen using the plurality method followed by a runoff between the first- and second-place finishers?

12. Eleven girls on a basketball team want to choose the type of food to be served at a team party. Each of the girls has an order of preference for the different types of food. The preference rankings of the girls are listed in the following table.

Number of Voters

	3	1	1	1	3	2
Pizza	1	2	3	4	2	3
Hamburgers	2	1	1	3	3	2
Hot dogs	3	3	2	1	4	4
Chicken	4	4	4	2	1	1

(a) Which type of food would win using the plurality method?

(b) Which type of food would win using the plurality method followed by a runoff between the top two finishers?

13. A poll of members of a community asked people about their opinions on a curfew time for young people under the age of 18. Their preference rankings broke down into the following percentages.

Percentage of Voters

	9	2	6	20	4	17	10	32
10 P.M.	1	2	3	4	3	4	4	4
12 midnight	2	1	1	1	2	2	3	3
1 A.M.	3	3	2	2	1	1	1	2
No curfew	4	4	4	3	4	3	2	1

(a) Which option would win using the plurality method?

(b) Which option would win using the plurality method followed by a runoff between the top two finishers?

14. Suppose three candidates—Whitney, Pronzini, and O'Kane—are running for mayor, and the preference rankings of the voters, broken down into percentages, are given in the following table.

Percentage of Voters

	21	14	20	19	16	10
Whitney	1	1	2	3	2	3
Pronzini	2	3	1	1	3	2
O'Kane	3	2	3	2	1	1

(a) In a plurality election, who would win?

(b) In a plurality election with a runoff between the top two finishers, who would win?

15. A college sorority must decide how many hours of community service should be required of its members each year. The preference rankings of the members are listed here.

Number of Voters

	3	1	1	1	2	1	2	8
0 hours	1	2	3	3	5	3	5	5
5 hours	2	1	1	2	3	2	4	4
10 hours	3	3	2	1	1	1	3	3
20 hours	4	4	4	4	2	4	1	2
40 hours	5	5	5	5	4	5	2	1

(a) Which choice would be selected in a plurality election?

(b) Which choice would be selected in a plurality election with a runoff between the top two finishers?

(c) In a plurality election, could the three members who ranked 0 hours first achieve a preferable outcome by voting strategically if the other members voted as shown in the table?

(d) In a plurality election with a runoff between the top two finishers, could the three members who ranked 0 hours first achieve a preferable outcome by voting strategically if the other members voted as shown in the table?

16. Three candidates are running for president of a service organization. The preference rankings of the voters are as follows:

Number of Voters

	13	5	10	7	10	6
Rosenthal	1	1	2	3	2	3
Salinas	2	3	1	1	3	2
Milner	3	2	3	2	1	1

(a) Who would win in a plurality election?

(b) Who would win in a plurality election with a runoff between the top two finishers?

(c) In a plurality election, could the six voters who ranked Milner first and Salinas second achieve a preferable outcome by voting strategically if the others voted as shown in the table?

(d) In a plurality election with a runoff between the top two finishers, could the six voters who ranked Milner first and Salinas second achieve a preferable outcome by voting strategically if the others voted as shown in the table?

Exercises 17–26 demonstrate how critical the method or order of voting may be to the outcome. In each of these questions we consider the situation where 12 people at a picnic want to select an activity for the afternoon. Their preference rankings of the possibilities are listed here.

Number of Voters

	1	1	1	1	3	2	2	1
Football	1	1	2	2	3	2	3	4
Soccer	2	4	1	4	2	4	2	3
Softball	3	2	3	1	1	3	4	2
Volleyball	4	3	4	3	4	1	1	1

17. (a) Which activity would win a plurality election?
    (b) In a plurality election, could the two whose first choice was football have achieved a preferable outcome by voting strategically if the other ten voted as shown in the table?
18. If the group did not have volleyball equipment, which of the remaining choices would win a plurality of the votes?
19. If the group did not have enough gloves for softball, which of the other three choices would win a plurality of the votes?
20. (a) In a plurality election with a runoff between the top two finishers, which activity would be selected?
    (b) In a plurality election with a runoff between the top two finishers, could the two voters who ranked volleyball first, soccer second, football third, and softball last achieve a better outcome by voting strategically if the other ten voted as shown in the table?
21. If the vote for activity is sequential, where the picnickers decide between football and soccer, then between the winner of that vote and softball, then between the winner of that vote and volleyball, which activity will they choose (if it is not dark by the time they decide)?
22. If the vote for activity is sequential, where the picnickers decide between softball and volleyball, then between the winner of that vote and soccer, then between the winner of that vote and football, which activity will they choose?
23. Is it possible to determine a sequence of votes, such as those described in Exercises 21 and 22, that yields soccer as the winner?
24. Suppose the vote is tournament style, with football matched against soccer and softball matched against volleyball, with a vote between winners determining the activity. Which activity wins?
25. Suppose the vote is tournament style, first with football matched against volleyball and soccer matched against softball, followed by a vote between the two winners. Which activity wins?
26. For each of the four activities, is it possible to arrange a tournament as in Exercises 24 and 25 so that activity wins?

1.2 BORDA'S METHOD: A SCORING SYSTEM

Elections determined by the plurality method or the plurality method with a runoff do not take into account the voters' relative preferences for *all* the candidates. A voter's top choice among the alternatives is revealed in a plurality election, but the balloting reveals nothing about the voter's last choice. In analyzing plurality methods with or without runoff, we used the assumption that we knew the preference rankings of the voters to determine which candidate particular voters would support in a given round of voting. However, in plurality elections the voters are not asked to give their preference rankings. In this and the following section, we assume that voters are asked to list their full set of preferences on a ballot, and we look at methods that use all this information.

One method of voting that requires voters to list their preference ranking for all choices is a scoring system called Borda's method, named for the Frenchman Jean-Charles de Borda (1733–1799). However, the method goes back at least to the second century A.D. Roman Senate. The method works as follows:

Borda's Method

With **Borda's method**, voters rank the entire list of candidates or choices in order of preference from their first choice to their last choice. After the votes have been cast, they are tallied as follows: On a particular ballot, the lowest-ranked candidate is given 1 point, the second lowest is given 2 points, and so on, to the top candidate who receives points equal to the number of candidates. The number of points given each candidate is summed across all ballots. We call this total the **Borda count** for the candidate. The winner is the candidate with the highest Borda count.

This method is illustrated in the following example.

EXAMPLE 1 *Borda's Method*

Suppose a five-member committee needs to select a chair from among three candidates named Coleman, Horowitz, and Taylor. They decide to use Borda's method. The preference rankings of the five committee members are recorded in Table 1.5.

TABLE 1.5 Preference Rankings for Committee Chair

	Number of Voters			
	1	2	1	1
Coleman	1	2	2	3
Horowitz	2	1	3	2
Taylor	3	3	1	1

Who will be the winner using Borda's method?

SOLUTION:

The Borda count for Coleman is given by

(no. of 1st-place votes)3 + (no. of 2nd-place votes)2 + (no. of 3rd-place votes)1
$$= 1 \cdot 3 + 3 \cdot 2 + 1 \cdot 1 = 3 + 6 + 1 = 10.$$

Similarly, the Borda count for Horowitz is

$$2 \cdot 3 + 2 \cdot 2 + 1 \cdot 1 = 6 + 4 + 1 = 11$$

and the Borda count for Taylor is

$$2 \cdot 3 + 0 \cdot 2 + 3 \cdot 1 = 6 + 0 + 3 = 9.$$

We see that Horowitz is the winner of the election. ■

TECHNOLOGY *tip*

A simple spreadsheet can do all the arithmetic required by Borda's method given the voters' preference rankings.

There are other equivalent ways to determine the Borda winner. One way is to give 0 points to the person last in the voter's ranking, 1 to the next-to-last person, and so on, up to the top-ranked person who receives points equal to one less than the number of candidates. The Borda method can also be carried out by giving 1 point to the top-ranked candidate, 2 to the candidate ranked second, and so forth, the Borda winner being the candidate with the lowest point total.

Notice that in Example 1, if a plurality election had been held, the result would have been a tie between Horowitz and Taylor. If a runoff between Horowitz and Taylor were held, we would expect their supporters to vote for them again, and we would expect the Coleman supporter to vote for the candidate that he or she ranked second. This would result in a runoff win for Horowitz, and we see that a plurality with a runoff and Borda's method give the same result. However, these two methods do not always result in the same winner. In fact, as we see in the next example, the Borda's method winner may not even make it to the runoff election.

EXAMPLE 2 *Comparison of Voting Methods*

Suppose that, in a survey, people were asked to rank the ice cream flavors chocolate, vanilla, and strawberry in order from their first to last choice, with the results given in Table 1.6.

TABLE 1.6 Ice Cream Preference Rankings

| | *Percentage of Voters* | | | | | |
	33	3	10	20	7	27
Chocolate	1	1	2	3	2	3
Vanilla	2	3	1	1	3	2
Strawberry	3	2	3	2	1	1

(a) Which flavor would win using Borda's method?
(b) Which flavor would win using the plurality method?
(c) Which flavor would win using the plurality method with a runoff between the first- and second-place finishers in the plurality election?

SOLUTION:

(a) The voting data are given in the form of percentages, but we can simply treat the percentages as we would voters and apply Borda's method as usual. The Borda count for chocolate is

(pct. of 1st-place votes)3 + (pct. of 2nd-place votes)2 + (pct. of 3rd-place votes)1
$$= (33 + 3) \cdot 3 + (10 + 7) \cdot 2 + (20 + 27) \cdot 1 = 36 \cdot 3 + 17 \cdot 2 + 47 \cdot 1 = 189.$$

Similarly, the Borda count for vanilla is

$$30 \cdot 3 + 60 \cdot 2 + 10 \cdot 1 = 220$$

and the Borda count for strawberry is

$$34 \cdot 3 + 23 \cdot 2 + 43 \cdot 1 = 191.$$

We see that the winner using Borda's method is vanilla.

(b) In a plurality election, the people vote for the flavor they rank as a first choice, so the results of the election would be

chocolate	36%
vanilla	30%
strawberry	34%

The winning flavor in a plurality election is chocolate. We see that vanilla, the winner using Borda's method, comes in last under the plurality method.

(c) Because chocolate and strawberry were the first- and second-place winners in the plurality election, they will be the two available choices in the runoff election. Notice that vanilla, the Borda's method winner, did not even make it into the runoff. In the runoff, the people who ranked vanilla first will now vote for their second-choice flavor. Tabulating these numbers, we arrive at the following runoff election result:

chocolate   36% + 10% = 46%
strawberry   34% + 20% = 54%

and we see that the winner in this case is strawberry.   ■

Just as with the plurality and plurality with runoff methods, with Borda's method voters may be able to get a preferable result by voting strategically. This phenomenon is illustrated in the next example in the context of Borda's method.

EXAMPLE 3 *Borda's Method and Strategic Voting*

Ten members of the families of a couple planning a wedding want to decide what kind of music will be played at the wedding reception. They have narrowed down the choices to country, jazz, polka, or rock and roll. For family harmony, they decide to vote using Borda's method, and the results are in Table 1.7.

TABLE 1.7 Wedding Music Preference Rankings

	Number of Voters							
	1	1	1	3	1	1	1	1
Country	1	1	1	2	2	3	4	4
Jazz	2	3	3	4	4	2	2	3
Polka	3	2	4	1	3	4	3	2
Rock and roll	4	4	2	3	1	1	1	1

(a) Which musical style will win using Borda's method?
(b) Can the individual who ranked the music in the order rock and roll first, country second, polka third, and jazz fourth vote strategically in such a way that the Borda winner would be rock and roll if the others voted as shown in the table?
(c) The two who ranked country music last are probably quite disappointed, particularly if they are the bride and groom. Could they have achieved a preferable outcome by voting strategically if the other eight voted as shown in the table?

SOLUTION:

(a) Country music received 3 first-place votes, 4 second-place votes, 1 third-place vote, and 2 fourth-place votes, so its Borda count is

$$3 \cdot 4 + 4 \cdot 3 + 1 \cdot 2 + 2 \cdot 1 = 28.$$

We compute the other Borda counts similarly, and the final tally is country 28, jazz 19, polka 26, and rock and roll 27. Therefore, country music is the Borda winner.

(b) Suppose the individual who ranked rock and roll first, country second, polka third, and jazz last instead ranked rock and roll first, jazz second, polka third, and country last. To see how this would affect the outcome of the election, we could completely recompute all of the Borda counts. However, it is easier to start with the original counts from part (a) and find the new Borda counts by adding or subtracting the appropriate amount from the original counts. For instance, by putting country last rather than second, the strategic voter would reduce its Borda count by 2, so country music's Borda count would be 28 – 2 = 26. Similarly, by moving jazz from fourth place to second place, the strategic voter would increase jazz's count by 2, resulting in a count of 19 + 2 = 21. Because polka and rock and roll remained in the same places in the strategic voter's rankings, their Borda counts of 26 and 27, respectively, would remain the same. Therefore, we see that the final tally would be country 26, jazz 21, polka 26, and rock and roll 27, resulting in a win for rock and roll.

(c) Suppose both of the voters ranking country music last had ranked polka first, rock and roll second, jazz third, and country last. We could recompute all of the Borda counts to see what happens, but we again compute the new Borda counts by adjusting the original counts accordingly. Because neither of the strategic voters would have changed their ranking of country last, country's original Borda count of 28 would be unchanged. Because both voters moved rock and roll from first to second place, each of them would reduce rock and roll's counts by 1, resulting in a new Borda count of 27 – 2 · 1 = 25. By ranking polka first, one of the strategic voters moved it up one place, whereas the other moved it up two places. Therefore, polka's new Borda count would be 26 + 1 + 2 = 29. Jazz's ranking was moved down one by one of the strategic voters and was unchanged by the other, so its new Borda count would be 19 – 1 = 18. Therefore, the new Borda count would be country 28, jazz 18, polka 29, and rock and roll 25, resulting in a win for polka. Note that although both of the strategic voters would prefer rock and roll to polka, they cannot lower country's count or raise rock and roll's, so they must settle for the smaller improvement from country to polka. ■

In using Borda's method, there are so many computations that you may want to check your results. Of course, one way to check is to redo the calculation, but there is also an easy partial check. Suppose the election has c candidates and v voters. Then each voter casts c numerical votes totaling

$$1 + 2 + \cdots + c = \frac{c(c + 1)}{2}$$

Therefore, the sum of all the Borda counts will total $vc(c + 1)/2$.

A CLOSER *look*

SUMMING THE FIRST *n* INTEGERS

To arrive at the formula for the sum of all the Borda counts, we used the fact that the sum of the first n integers is $n(n+1)/2$. Let us see why this formula is true by looking at the sum of the first four integers. First we write this sum as $1 + 2 + 3 + 4$ and then in the reverse order as $4 + 3 + 2 + 1$. Then we add these together and see by the arithmetic and the corresponding picture that $2(1 + 2 + 3 + 4) = 4 \cdot 5$, or $1 + 2 + 3 + 4 = \dfrac{4 \cdot 5}{2}$.

The same argument shows that

$$1 + 2 + \cdots + n = \frac{n(n+1)}{2}.$$

$$
\begin{array}{l}
1 + 2 + 3 + 4 \\
4 + 3 + 2 + 1 \\
\hline
5 + 5 + 5 + 5 = 4 \cdot 5
\end{array}
$$

In a Borda's method election with c candidates and v voters, the sum of all the Borda counts will be

$$\frac{vc(c + 1)}{2}.$$

If the election results are given in the form of percentages, we set the number of voters, v, in the above formula equal to 100.

Sometimes we would like to use Borda's method, but the number of candidates or choices the voters must rank is rather overwhelming. In this case, a modified version of Borda's method may be used in which the voters rank only their top few choices. The voting method for the Heisman Memorial Trophy Award is of this type. On the Heisman ballot, voters are asked to rank only their top three choices from among all college football players in the United States. The Borda count for each candidate is computed by giving 3 points for each first-place vote, 2 points for each second-place vote, and 1 point for each third-place vote. Any candidate not listed on a ballot will get 0 points from that ballot. The winner is declared to be the candidate with the highest Borda count. In sports polls where this form of voting is commonly used, the voters may know a lot about the top teams or players and be able to rank them, but may not know enough to rank all eligible candidates, so lumping all but the top candidates together with 0 points simplifies the process for the voters.

Borda's method is just one voting method that takes into account the entire preference rankings of voters. Some methods that are similar to Borda's method assign a wider spread of points to the different rankings. For instance, in voting for the National Basketball Association Most Valuable Player award, over one hundred members of the media list their first through fifth choices for the award. Each first-place vote receives 10 points, each second-place vote 7 points, each third-place vote 5 points, each fourth-place vote 3 points, and each fifth-place vote 1 point.

branching OUT

DETERMINING WHO IS NUMBER ONE IN COLLEGE FOOTBALL

TCU players Marcus Cannon (61), Andy Dalton (14), and Josh Vernon (78) at the 2011 Rose Bowl.

In 1936, Alan Gould, the sports editor of the Associated Press (AP), came up with the idea of the AP college football poll to satisfy his customers—the subscriber newspapers of the Associated Press. The newspapers were looking for a way to arouse interest in, and controversy about, college football. Gould's idea of a national poll fit the bill. The voters in the AP poll are newspaper, radio, and television sports reporters throughout the country.

In 1950, United Press International started a second college football poll, in which coaches, rather than media people, were the voters. This poll is now sponsored by *USA Today*. In 2010, there were 59 coaches with votes. The rankings are determined by a Borda-type scheme. Each voter ranks his choice for the top 25 football teams in the NCAA Football Bowl Subdivision in order. A first-place vote is worth 25 points, a second-place vote is worth 24 points, and so on, down to a 25th-place vote, which is worth 1 point. There are 120 teams in the NCAA Football Bowl Subdivision, and each voter can choose the top 25 from among any of these 120 teams. Teams not

USA Today Top 25 in College Football, January 11, 2011 (First-place votes are in parentheses.)

	Record	Points
1. Auburn (56)	14-0	1424
2. TCU (1)	13-0	1336
3. Oregon	12-1	1333
4. Stanford	12-1	1254
5. Ohio State	12-1	1197
6. Oklahoma	12-2	1096
7. Boise State	12-1	1012
8. LSU	11-2	1007
9. Wisconsin	11-2	1007
10. Oklahoma State	11-2	883
11. Alabama	10-3	860
12. Arkansas	10-3	818
13. Nevada	13-1	734
14. Michigan State	11-2	676
15. Virginia Tech	11-3	636
16. Florida State	10-4	506
17. Mississippi State	9-4	505
18. Missouri	10-3	473
19. Nebraska	10-4	354
20. UCF	11-3	328
21. Texas A&M	10-4	277
22. South Carolina	9-5	181
23. Utah	10-3	156
24. Maryland	9-4	111
25. North Carolina State	9-4	94

Others receiving votes: Northern Illinois 82, Tulsa 41, San Diego State 36, West Virginia 35, Iowa 31, Miami (Ohio) 13, Florida 10, Connecticut 7, Air Force 4, Hawaii 4, Notre Dame 3, Washington 1.

included in a voter's ranking get no points from that voter. The final ranking of the teams is determined by the teams' point total across all of the voters. The results are published in newspapers across the country. The final *USA Today* Poll from the 2010 season is shown in the table.

Unlike most collegiate sports, the NCAA Football Bowl Subdivision does not have a playoff to determine a national champion. Beginning with the 1998 season, the teams ranked number 1 and 2 at the end of the regular season according to the BCS (Bowl Championship Series) formula that combines human polls and computer rankings at the end of the regular season have automatically played each other in the BCS National Championship Game. The precise BCS formula has changed over the years. The formula in 2010 was an equally weighted average of the *USA Today* Coaches Poll; a similar poll, the Harris Interactive Poll; and the average of six computer rankings. Because millions of dollars ride on the possibility of playing in the championship game, the *USA Today* and Harris polls are quite important. This leads to the ethical question of whether or not the football rankings should be affected by the votes of coaches and media people who may have a vested interest in particular teams. Of particular concern is the temptation to manipulate the results by use of strategic voting. Because a Borda-type method is used in the polls, a single voter has a great deal of power to affect the outcome. For instance, if a voter's favored team is in a close contest for first place with another team, the voter can damage the other team's prospects significantly by ranking them very low on the ballot or not even ranking them all. It is unlikely that the controversy surrounding the national championship of college football will be resolved by anything other than the institution of a playoff that includes all teams that might reasonably be considered in the running for the title.

The advantage of Borda's method over plurality methods is that voters are able to express their opinions about candidates other than just their first choice candidate. Because of this, a candidate rated highly, but not necessarily first, by most voters will often win an election decided by Borda's method. However, this advantage is somewhat offset by the fact that Borda's method is more susceptible to manipulation by strategic voters than are the plurality and plurality with runoff methods.

EXERCISES FOR SECTION 1.2

1. A 14-member board used Borda's method to elect a chair. The four candidates were Cardona, Pitts-Jones, De Plata, and Vincent. The preference rankings on the 14 ballots are listed here.

Number of Voters

	1	4	1	2	2	1	1	2
Cardona	1	1	1	3	4	2	2	3
Pitts-Jones	2	3	3	1	1	3	4	2
De Plata	3	2	4	2	2	1	1	4
Vincent	4	4	2	4	3	4	3	1

Who was the winner using Borda's method?

2. The returning members of a basketball team decide to select a team captain for the next season using Borda's method. The three candidates are Thomas, Walker, and Goodman. The preference rankings on the nine ballots are listed here.

	Number of Voters					
	3	1	1	1	2	1
Thomas	1	1	2	3	2	3
Walker	2	3	1	1	3	2
Goodman	3	2	3	2	1	1

Who is the winner using Borda's method?

3. An eight-member committee must decide on a school mascot. The choices have been narrowed down to the following: Bears, Falcons, Trojans, Mustangs, and Horned Frogs. The preference rankings of the committee members are listed here.

	Number of Voters							
	1	1	1	1	1	1	1	1
Bears	1	2	3	2	4	3	5	5
Falcons	3	1	1	4	3	2	4	4
Trojans	5	4	5	1	1	4	2	3
Mustangs	4	5	4	5	5	1	3	2
Horned Frogs	2	3	2	3	2	5	1	1

Which mascot is the winner using Borda's method?

4. A publishing company plans to open a new office in one of the following six cities: Chicago, New York, Philadelphia, Seattle, Los Angeles, or Denver. An 11-member executive board must decide on the city. The preference rankings of each board member are given next.

	Number of Voters										
	1	1	1	1	1	1	1	1	1	1	1
Chicago	4	5	5	3	3	3	5	2	2	3	4
New York	1	1	1	6	5	6	4	4	4	2	3
Philadelphia	3	4	3	1	2	2	2	5	5	5	6
Seattle	6	3	4	4	4	5	3	3	6	6	2
Los Angeles	5	6	2	2	1	1	1	6	3	4	5
Denver	2	2	6	5	6	4	6	1	1	1	1

Which city is the winner if Borda's method is used?

5. A small company decided to use Borda's method to determine on what day of the week to have a company picnic. The three choices were Monday, Wednesday, and Thursday. Employees ranked their choices as shown here.

Number of Voters

	8	10	8	7	4	8
Monday	1	1	2	3	2	3
Wednesday	2	3	1	1	3	2
Thursday	3	2	3	2	1	1

Which day won using Borda's method?

6. Sixty-four people were registered to attend a conference. The person in charge of planning the conference decided to use Borda's method to let the participants vote on a group recreational activity. The four choices were ballet, opera, baseball game, and horse racing. The preference rankings of the people are given here.

Number of Voters

	7	14	5	6	7	5	10	10
Ballet	1	1	2	2	3	4	3	4
Opera	2	2	1	1	4	3	4	3
Baseball game	3	4	3	4	1	1	2	2
Horse racing	4	3	4	3	2	2	1	1

Which activity won using Borda's method?

7. Three candidates—Kahn, Papini, and Tilson—ran for a seat on city council. Polls indicate that the preference rankings of the voters broke down into the following percentages.

Percentage of Voters

	26	13	5	22	18	16
Kahn	1	1	2	3	2	3
Papini	2	3	1	1	3	2
Tilson	3	2	3	2	1	1

Who would have been the winner of the election if Borda's method had been used?

8. Candidates Mijares, Zapata, and Schwartz are running for president of a professional society. The election is to be decided using Borda's method, and the preference rankings of the voters break down into the following percentages.

Percentage of Voters

	17	14	22	15	17	15
Mijares	1	1	2	3	2	3
Zapata	2	3	1	1	3	2
Schwartz	3	2	3	2	1	1

Who is the winner using Borda's method?

9. The preference rankings of a class of 22 college students for their favorite fictional news anchor are as follows:

	Number of Voters					
	3	3	6	2	5	3
Tom Grunick	1	1	2	3	2	3
Ron Burgundy	2	3	1	1	3	2
Hector Madden	3	2	3	2	1	1

   (a) Which anchor has the top Borda count?
   (b) Which anchor wins a plurality of the vote?
   (c) Which anchor wins using the plurality method with a runoff between the top two finishers?

10. An engineering team needed to select a team leader from among the candidates Tate, Kummer, Dobbins, and Coscia. Dobbins was not well liked and was ranked last by everyone but himself. Of course, he ranked himself first. The preference rankings of the members were as follows:

	Number of Voters						
	7	3	8	4	1	4	3
Tate	1	1	2	3	2	2	3
Kummer	2	3	1	1	4	3	2
Dobbins	4	4	4	4	1	4	4
Coscia	3	2	3	2	3	1	1

   (a) Which candidate would have won using Borda's method?
   (b) Which candidate would have won using the plurality method?
   (c) Which candidate would have won using the plurality method with a runoff between the first- and second-place finishers?

11. A 25-member division needs to decide which sport to play at their picnic. The preference rankings of the division members follow:

	Number of Voters					
	4	3	5	3	6	4
Soccer	1	1	2	3	2	3
Baseball	2	3	1	1	3	2
Football	3	2	3	2	1	1

   (a) Which sport wins using the plurality method?
   (b) Which sport wins using the plurality method with a runoff between the first- and second-place finishers?
   (c) Which sport wins using Borda's method?
   (d) If Borda's method is used, can the four voters who ranked soccer first, baseball second, and football third obtain a preferable result by voting strategically if the others vote as shown in the table?

12. A soccer team wants to select a color for their uniforms. The preference rankings of the team members are listed in the following table.

Number of Voters

	4	3	5	2	3	1
Blue	1	1	2	3	2	3
Green	2	3	1	1	3	2
Red	3	2	3	2	1	1

(a) Which color wins using the plurality method?

(b) Which color wins using the plurality method with a runoff between the top two finishers?

(c) Which color wins using Borda's method?

(d) Could the members whose first choice was green obtain a preferable result in an election decided by Borda's method by voting strategically if the other members voted as shown in the table?

13. An English department with ten members is selecting a new chair. Six members of the department are eligible and willing to run for the position. The preference rankings of the department members are given next.

Number of Voters

	1	1	2	1	1	1	1	2
LeBreton	1	2	6	6	4	5	5	3
Frye	5	1	1	1	5	4	6	6
Reeves	2	6	4	2	1	6	2	4
Keen-Sims	4	5	3	5	6	1	1	5
Cullors	6	3	2	3	2	3	3	2
Allen	3	4	5	4	3	2	4	1

(a) Who has the top Borda count?

(b) If Borda's method is used, could the members who ranked Allen first obtain a preferable result by voting strategically if the other members voted as shown in the table?

(c) If Borda's method is used, could the member who ranked Keen-Sims first and Reeves second obtain a preferable result by voting strategically if the other members voted as shown in the table?

14. A PTA board voted to select how to spend $20,000 they had earned in a fundraiser. The preference rankings of the voting members appear next.

Number of Voters

	4	3	1	2	2	1	1	6	2
Computers	1	1	2	2	3	3	3	2	3
Library books	2	3	1	1	1	2	4	3	2
Playground equipment	4	4	3	4	4	1	1	4	4
Science lab	3	2	4	3	2	4	2	1	1

(a) Which choice has the top Borda count?

(b) If Borda's method is used, could the member who ranked playground equipment first and science lab second obtain a preferable result by voting strategically if the other members voted as shown in the table?

(c) If Borda's method is used, could the two members who ranked library books first, science lab second, computers third, and playground equipment last obtain a preferable result by voting strategically if the other members voted as shown in the table?

Exercises 15 and 16 consider two of the several alternate definitions of Borda's method.

15. Show that when, for each preference ballot, the lowest-ranked candidate is given 0 points, the second lowest is given 1 point, and so on, to the top candidate who receives points one less than the number of candidates, and the results are summed across all ballots, the candidate with the most points is always the winner of the Borda count according to the definition in the text.

16. Show that when, for each preference ballot, the top candidate is given 1 point, the second highest candidate 2 points, and so forth, and the results are summed across all ballots, the candidate with the fewest points is always the winner of the Borda count according to the definition in the text.

1.3 HEAD-TO-HEAD COMPARISONS

If we know the preference rankings of the voters, as we must to apply Borda's method, we can also see who would be the winner were any two of the candidates pitted against each other in a plurality election in the absence of the other candidates. We call such a plurality election between any two of the candidates a **head-to-head comparison**. To see how this is done, we refer back to an earlier example.

EXAMPLE 1 *Head-to-Head Comparisons*

In the election between Coleman, Horowitz, and Taylor that we looked at in Example 1 of the previous section, the preference rankings were listed as in Table 1.8.

TABLE 1.8 Preference Rankings for Committee Chair

	Number of Voters			
	1	2	1	1
Coleman	1	2	2	3
Horowitz	2	1	3	2
Taylor	3	3	1	1

(a) Who would be the winner in a head-to-head comparison between Coleman and Horowitz?

(b) Who would be the winner in a head-to-head comparison between Coleman and Taylor?

(c) Who would be the winner in a head-to-head comparison between Horowitz and Taylor?

SOLUTION:

(a) In Table 1.9 those ballots on which Coleman is ranked above Horowitz are shaded blue. These two voters would vote for Coleman in a head-to-head comparison with Horowitz. The ballots on which Horowitz is ranked above Coleman are shaded pink. These three voters would vote for Horowitz in a head-to-head comparison between Coleman and Horowitz. Therefore, Horowitz would win with 3 votes to Coleman's 2 votes in a head-to-head comparison.

TABLE 1.9 Head-to-Head Comparison Between Coleman and Horowitz

	Number of Voters			
	1	2	1	1
Coleman	1	2	2	3
Horowitz	2	1	3	2
Taylor	3	3	1	1

(b) Again, by comparing the rankings of Coleman and Taylor on each of the ballots, we see that Coleman would win in a head-to-head comparison with Taylor by receiving 3 votes, whereas Taylor would receive 2 votes.

(c) From the rankings, we see that Horowitz would receive 3 votes in a head-to-head comparison with Taylor, whereas Taylor would receive the remaining 2 votes, so Horowitz would win. ∎

In Example 1 we saw that Horowitz beat both of the other candidates in head-to-head comparisons. Horowitz is an example of a Condorcet winner, named after the Frenchman Marie Jean Antoine Nicolas Caritat, the Marquis de Condorcet (1743–1794), and defined as follows.

Condorcet Winner

A candidate who is the winner of a head-to-head comparison with every other candidate is called a **Condorcet winner**. If a candidate beats or ties every other candidate in head-to-head comparisons, we call that candidate a **weak Condorcet winner**. A given election may or may not have a Condorcet winner or a weak Condorcet winner.

As with Borda's method, the first to propose the method is unclear. It goes back at least as far as the Catalonian Ramon Llull in 1299. In *The Art of Elections*, Llull advocated selecting the Condorcet winner in elections within the Catholic Church. Many others have agreed that a Condorcet winner deserves to be elected. Charles Dodgson, better known to us as Lewis Carroll, the author of *Alice's Adventures in Wonderland*, was a lecturer in mathematics at Christ Church College, Oxford. He supported Condorcet's view as did the Welsh economist Duncan Black, whose 1963 book, *The Theory of Committees and Elections*, was one of the pioneering works on voting methods.

TECHNOLOGY *tip*

A spreadsheet or computer program can be set up to make all of the head-to-head comparisons required to determine a Condorcet winner, given the voters' preference rankings.

Although many view the election of a Condorcet winner as desirable and some view it as mandatory, a problem remains. There may be no Condorcet winner!

EXAMPLE 2 *An Election with No Condorcet Winner*

This example originated with Condorcet in 1785. The preference rankings of three voters deciding between candidates A, B, and C are listed in Table 1.10.

TABLE 1.10 Preference Rankings Without a Condorcet Winner

	Number of Voters		
	1	**1**	**1**
A	1	3	2
B	2	1	3
C	3	2	1

Show that there is no Condorcet winner.

SOLUTION:

In a head-to-head match-up between candidates A and B, the rankings show that A beats B by a score of 2 to 1. Similarly, C beats A by a score of 2 to 1, and B wins over C by a score of 2 to 1. We see that no candidate beats every other candidate in head-to-head match-ups, and therefore there is no Condorcet winner. ∎

In Examples 1 and 2, we had the preference rankings of all the voters and hence we could find the Condorcet winner or determine that there was none. Preference rankings cannot be completely worked out from the results of most elections, but combining information about the nature of the candidates with information gathered from exit polls may allow us to draw reasonable conclusions about the voters' preferences and hence about whether or not there is a Condorcet winner. Hypothesizing preference rankings from limited data is an example of *mathematical modeling*, in which real-life problems are analyzed by deducing or hypothesizing a mathematical structure. This is an important problem-solving strategy. The validity of a model can be measured by how well it agrees with real-life observations. We have more confidence in the predictions based on a model if they are not significantly affected by small changes in the model's most debatable assumptions. As an example of this type of analysis, we consider the 1990 election for governor of Louisiana.

EXAMPLE 3   *1990 Louisiana Gubernatorial Election*

The three main candidates in the 1990 election for governor of Louisiana were David Duke, formerly a Grand Wizard in the Ku Klux Klan; Edwin Edwards, ex-governor, who at the time had twice been indicted but not convicted for fraud, (though in 2001 he was sentenced to 10 years in prison for racketeering); and incumbent Buddy Roemer, whose switch from the Democratic to the Republican Party and efforts to reform state government alienated many voters. The election was conducted using the plurality method followed by a runoff election between the top two finishers. On the initial ballot, Edwards won 34% of the vote, Duke 32%, and Roemer 27%, with the remaining 7% going to other candidates. After receiving a grudging endorsement from Roemer for the runoff, Edwards prevailed over Duke 60% to 40%.

Duke, Edwards, and Roemer (left to right).

To guess at a reasonable distribution of preference rankings or second choices, we begin by scaling the initial vote to 100%. Because the initial percentages for the three candidates total 93%, we multiply each candidate's share of the vote by 100/93. After rounding, this gives us Edwards 37%, Duke 34%, and Roemer 29%.

Because Edwards only received 37% of the (adjusted) initial vote and he received 60% of the vote in the runoff, it is reasonable to suppose that 60% − 37% = 23% of the voters originally supported Roemer but switched to Edwards in the runoff. These voters would rank the candidates in the order Roemer, Edwards, and Duke from first to last choice.

Similarly, because Duke received 34% in the initial vote and 40% in the runoff, we would suppose that 6% of the voters supported Roemer originally and switched to Duke in the runoff. These voters would rank the candidates in the order Roemer first, Duke second, and Edwards last.

We can only guess the second and third choices for Edwards and Duke supporters. In light of the split of the Roemer supporters and of the candidates' political baggage, one approximation of the breakdown of preference rankings is given in Table 1.11.

TABLE 1.11 Possible Preference Rankings for the 1990 Louisiana Governor's Election

	Percentage of Voters					
	9	28	11	23	23	6
Edwards	1	1	2	3	2	3
Duke	2	3	1	1	3	2
Roemer	3	2	3	2	1	1

Based on this table, who is the Condorcet winner?

SOLUTION:

Using our figures, in a head-to-head comparison of Roemer with Edwards, Edwards would have his 37% plus the 11% who ranked him second behind Duke, giving him 37% + 11% = 48% of the vote. Roemer would get the remaining 52% of the vote, and thus Roemer would have beaten Edwards 52% to 48%. Similarly, Roemer would defeat Duke 57% to 43%. Thus, Roemer is the Condorcet winner. ∎

Note that the conclusions in this example depended on the suppositions we made about the breakdown of voter rankings. We assumed about 67.6% of Duke's supporters ranked Roemer second. If instead we had assumed fewer than 62% of Duke's supporters had Roemer as a second choice, then Edwards would have been the winner in a head-to-head comparison with Roemer.

The situation in Example 3 shows that a Condorcet winner may not even make the runoff in a plurality with runoff election. It is also true that a Condorcet winner may not win an election under the Borda method. However, in arguing in favor of the Condorcet winner, one could regard the roles of all other candidates as spoilers, because none would beat the Condorcet winner in a head-to-head match-up. In horse racing, middle- to long-distance races in track meets, and bicycle races, it is common for a "rabbit" or pacer with no chance of winning to be entered in order to increase the chances of the contender on the "team." In many ways, an election or choice among several alternatives may be viewed in a similar light, even if the effect is unintentional.

How common is an election with no Condorcet winner? With three candidates and an odd number of voters (to avoid ties) who order the candidates at random, the probability that there is a Condorcet winner decreases as the number of voters increases, but is always above 91%. Similarly, the probability of a Condorcet winner when an odd number of voters randomly order four candidates exceeds 82%. For five and six candidates, the percentages exceed 74% and 68%, respectively. In real-life situations there are often some strong candidates and some weaker candidates, so the likelihood of there being a Condorcet winner is significantly higher.

In a race with c candidates, there are $c(c-1)/2$ head-to-head comparisons. Computing all of them could take a bit of time. A more efficient way to look for a Condorcet winner is to begin with two strong candidates, perhaps the first- and second-place finishers by the plurality method or Borda's method. The winner of this contest would be compared with a third candidate, the winner of this with a fourth, and so on. Once all candidates except one have been defeated in this way, the remaining candidate is faced off against candidates not yet compared with until this candidate loses or defeats all candidates. Proceeding in this way only requires from $c - 1$ to $2c - 3$ head-to-head comparisons, depending on the particular outcomes.

Single-Peaked Preference Rankings

Before deciding to use the Condorcet method, it would be nice to be assured that a Condorcet winner exists. There is a condition that guarantees a Condorcet winner when the number of voters is odd or a weak Condorcet winner when the number of voters is even. We now examine this condition in the case of an election with four candidates, A, B, C, and D. Suppose that the candidates can be ordered in a line, for instance in the order C, D, A, B, in such a way that in ranking the candidates every voter makes a first choice and then moves only outward along the line from this first choice in ranking the remaining candidates. Three examples of rankings of this sort are shown in Figure 1.1.

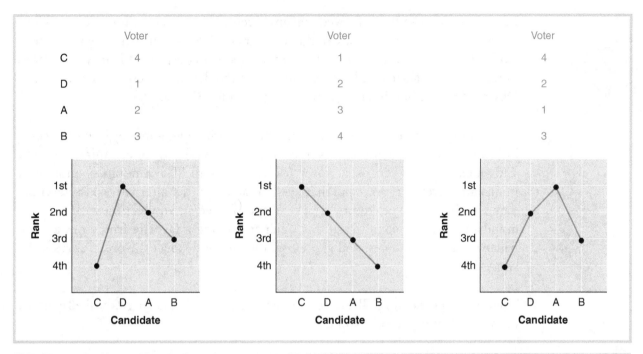

FIGURE 1.1 Rankings with a single peak.

Graphically, we see that these kinds of preference rankings have single peaks. For instance, the preferences of the blue voter have a single peak at candidate D, those of the green voter have a single peak at candidate C, and the preferences of the red voter have a single peak at Candidate A. In any election, if *all* voters' preference rankings have a single peak with respect to some *fixed* ordering of the candidates, we say that preference rankings are **single-peaked** with respect to that order. We cannot call the system of preferences single-peaked if even one voter gives a ranking that is not single-peaked with respect to the fixed order of the candidates, such as the ranking graphed in Figure 1.2, with peaks at Candidates C and A with respect to the order C, D, A, B. Although the graphs serve as motivation for the term single-peaked, we do not need to draw them to determine whether preferences are single-peaked. Once put in the order under consideration, all you need to do is make sure that as you move either up or down a voter's preferences from rank 1 the ranks always increase.

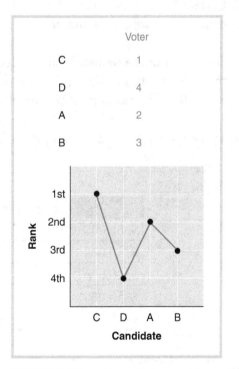

FIGURE 1.2 A ranking with two peaks.

Economist Duncan Black showed that any election in which the preference rankings are single-peaked has a weak Condorcet winner. The only time there will not be a Condorcet winner when preference rankings are single-peaked is if two or more candidates tie head-to-head against each other and defeat the rest of the candidates. Observe that the number of voters must be even in this case.

Preferences are often single-peaked in situations involving money, such as deciding the amount to commit to some project or cause, or determining a salary. This makes sense, because a voter probably has an ideal amount of money in mind and moves outward from that ideal in ranking other alternatives. Preferences may also be single-peaked or nearly single-peaked in an election with a single dominant issue. In some situations, such as ranking a favorite fruit from among bananas, apples, oranges, and grapes, you would not expect the preferences to be single-peaked.

The following example illustrates a situation in which the single-peaked pattern occurs, and, as expected, a Condorcet winner exists.

EXAMPLE 4 *Single-Peaked Preference Rankings and a Condorcet Winner*

Suppose a nine-member board is deciding the salary of a company's chief executive officer and has the options of $200,000, $250,000, $300,000, $350,000, and $400,000. The rankings of the members' choices of salaries are listed in Table 1.12.

TABLE 1.12 Preference Rankings for CEO's Salary

	Number of Voters			
	4	2	2	1
$200,000	1	4	5	5
$250,000	2	3	4	4
$300,000	3	1	1	3
$350,000	4	2	2	1
$400,000	5	5	3	2

Find the Condorcet winner.

SOLUTION:

Five of the nine board members favor $300,000 over $200,000 or $250,000, and eight of nine favor $300,000 over $350,000 or $400,000. Therefore, $300,000 is the Condorcet winner. Notice that the preferences are single-peaked with respect to the natural order $200,000, $250,000, $300,000, $350,000, $400,000. ■

An election procedure from which a Condorcet winner, if there is one, emerges victorious requires a back-up voting procedure to use in the case that no Condorcet winner exists. Such a process is called a **Condorcet completion process**. In 1876, Charles Dodgson suggested choosing the Condorcet winner if one exists and in the case of no Condorcet winner, he suggested several alternatives for deciding the election. In 1963, Welsh economist Duncan Black, in his study of voting, advocated using Borda's method when there is no Condorcet winner. Although feasible, this leaves the task of employing two completely independent methods to determine the result of the election. In the late nineteenth century, E. J. Nanson proposed using the Borda counts to eliminate all candidates with below average Borda counts. Voters' rankings of the remaining candidates are then used to compute new Borda counts, those with below average counts are eliminated, and the process continues until a single candidate is left. Exercise 20 asks you to prove that if there is a Condorcet winner, that candidate will be the winner under Nanson's method. Nanson's method has several variations. One is to eliminate only the lowest count before recomputing Borda counts. This has the advantage of being less subject to manipulation by a group of voters, but may require significantly more Borda count computations. This disadvantage is relatively minor when results are tallied by a computer.

A Condorcet winner has a good claim to being the most preferred choice in a multicandidate election and is largely unaffected by strategic voting or additional candidates without realistic prospects for winning. The drawback is that Condorcet winners do not always exist, so that any election procedure that elects a Condorcet winner, if one exists, must be augmented by a method to decide elections without one.

EXERCISES FOR SECTION 1.3

1. Three candidates—Bauer, Sanders, and Donevska—are running for president of a service organization with 12 members. The preference rankings of the members are listed here.

Number of Voters

	2	1	4	2	2	1
Bauer	1	1	2	3	2	3
Sanders	2	3	1	1	3	2
Donevska	3	2	3	2	1	1

    (a) Who is the winner in a head-to-head comparison between Bauer and Sanders?

    (b) Who is the winner in a head-to-head comparison between Bauer and Donevska?

    (c) Who is the winner in a head-to-head comparison between Donevska and Sanders?

    (d) Which candidate, if any, is the Condorcet winner?

2. Eight friends are trying to decide which movie to rent at the video store. They have narrowed the choices down to a comedy, western, or action movie. The preference rankings of the friends are listed here.

Number of Voters

	2	1	2	1	2
Comedy	1	2	3	2	3
Western	3	1	1	3	2
Action movie	2	3	2	1	1

    (a) Which is the winner in a head-to-head comparison between comedy and western?

    (b) Which is the winner in a head-to-head comparison between comedy and action movie?

    (c) Which is the winner in a head-to-head comparison between western and action movie?

    (d) Which type, if any, is the Condorcet winner?

3. The preference rankings of a class of college students for their favorite "ethnic" food are as follows:

Number of Voters

	1	1	2	1	1	1	1	2	1	2	3	3	2
Cajun	1	1	1	3	4	2	3	2	3	4	3	4	5
Chinese	2	3	4	1	1	4	4	4	2	2	4	3	3
Indian	3	5	5	5	5	5	5	5	5	5	5	5	4
Italian	4	4	3	2	2	1	1	3	4	3	2	2	2
Mexican	5	2	2	4	3	3	2	1	1	1	1	1	1

Which type of food, if any, is the Condorcet winner?

4. The members of a tennis team are asked to select the most valuable player on the team from among four candidates named Hall, Perez, Moore, and Chihara. The preference rankings of the members are as follows:

Number of Voters

	1	1	2	1	1	2	1
Hall	1	1	3	3	3	3	4
Perez	3	3	1	2	4	2	3
Moore	2	4	2	1	1	4	2
Chihara	4	2	4	4	2	1	1

Which player, if any, is the Condorcet winner?

5. The managers of a catering service meet to decide on their first ever employee-of-the-month. Three employees are suggested: Julia, Paul, and Wolfgang. The managers' preference rankings appear here.

Number of Voters

	1	1	3	2
Julia	1	1	3	2
Paul	2	3	1	3
Wolfgang	3	2	2	1

Which employee, if any, is the Condorcet winner?

6. Twelve members of a search committee in charge of hiring the new president for a group of banks have narrowed their list to three finalists: DeZern, Parker, and Schultz. After much discussion they decided to rank the finalists and offer the position to the Condorcet winner, if one exists. The rankings are given in the table.

Number of Voters

	2	2	3	1	3	1
DeZern	1	1	2	3	2	3
Parker	2	3	1	1	3	2
Schultz	3	2	3	2	1	1

Which candidate, if any, is the Condorcet winner?

7. A small-town newspaper asked its readers to rate their favorites in various categories. In the restaurant category, the choices were the Hungry Boar, the Hungry Dog, and the Hungry Pig. The percentage of those responding with each possible ranking is given in the table.

Percentage of Voters

	13	21	17	12	24	13
Hungry Boar	1	1	2	3	2	3
Hungry Dog	2	3	1	1	3	2
Hungry Pig	3	2	3	2	1	1

Which restaurant, if any, is the Condorcet winner?

8. An AM radio station in financial trouble is seeking a partner for a merger. Three potential partners have made offers. The AM station's stockholders must choose whether to merge with an FM radio station, a television station, or a newspaper. The percentage of stockholders with each possible preference ranking is given in the table.

	Percentage of Voters					
	16	17	20	9	11	27
FM radio station	1	1	2	3	2	3
Television station	2	3	1	1	3	2
Newspaper	3	2	3	2	1	1

Which option, if any, is the Condorcet winner?

9. The parents of children in an elementary school were asked to vote on having mandatory school uniforms, optional school uniforms, or no school uniforms. The preference rankings of the parents are as follows:

	Number of Voters					
	36	5	16	19	1	42
Mandatory uniforms	1	1	2	3	2	3
Optional uniforms	2	3	1	1	3	2
No uniforms	3	2	3	2	1	1

(a) Which option has the top Borda count?
(b) Which option would win a plurality of the vote?
(c) Which option would win using the plurality method with a runoff between the first- and second-place finishers?
(d) Which option, if any, is the Condorcet winner?

10. In order to promote a wider viewpoint, a board of directors of a company has decided to increase its size by adding from one to five members. The preference rankings regarding the number of members to add to the 11-member board are as follows:

	Number of Voters					
	2	1	2	1	1	4
One	1	3	5	5	5	5
Two	2	1	1	3	3	4
Three	3	2	2	1	2	3
Four	4	4	3	2	1	2
Five	5	5	4	4	4	1

(a) Which number has the top Borda count?
(b) Which number would win a plurality of the vote?

(c) Which number would win using the plurality method with a runoff between the first- and second-place finishers?

(d) Which number, if any, is the Condorcet winner?

11. A family of six is planning to buy a dog. They have narrowed the choice of breed down to four choices, and their preference rankings are as follows:

Number of Voters

	1	1	1	1	1	1
Beagle	1	1	1	3	3	4
Fox terrier	3	4	3	1	2	3
Bulldog	2	2	4	4	1	1
Boxer	4	3	2	2	4	2

(a) Which breed has the top Borda count?

(b) Which breed would win a plurality of the vote?

(c) Which breed would win using the plurality method with a runoff between the top two finishers?

(d) Which breed, if any, is the Condorcet winner?

12. Are the preference rankings in Exercise 10 single-peaked with respect to the ordering of candidates one, two, three, four, five? Would you expect them to be? Explain why or why not.

13. Are the preference rankings in Exercise 11 single-peaked with respect to the ordering of candidates beagle, fox terrier, bulldog, boxer?

14. Are the preference rankings in Exercise 11 single-peaked with respect to any ordering of the candidates? Justify your answer.

15. In January 1925, the U.S. Senate considered development of the Muscle Shoals area of the Tennessee River in Alabama. In successive votes the Senate voted for (1) private development over public development, 48-37; (2) recommitting the bill for further study over private development, 46-33; (3) public development over recommitment, 40-39; (4) private development over public development, 46-33. At this point the motion to recommit was again introduced. Because of White House lobbying, several senators switched votes and private development won over recommitment, 43-38. The bill finally passed in this version. Of the 96 members of the Senate, 75 voted on enough of these amendments and in a consistent manner on the first four votes so that we may surmise a preference ranking among the three alternatives. The results are as follows:

Number of Voters

	9	18	4	26	16	2
Private development	1	1	2	3	2	3
Public development	2	3	1	1	3	2
Recommitment	3	2	3	2	1	1

Source: Congressional Record.

Assume that only these 75 senators voted.

(a) Which alternative would win a plurality vote?

(b) Which alternative would win a runoff election between the top two finishers?

(c) Which alternative is the Borda winner?

(d) Which option, if any, is the Condorcet winner?

(The outcome in (d) was originally observed by John L. Neufeld, William J. Hausman, and Ronald B. Rapoport, under slightly different assumptions. See the Suggested Readings at the end of this chapter.)

16. The 1860 election, precursor to the Civil War, saw four candidates win states. Split along regional lines, the contest was primarily between Republican Abraham Lincoln and Democrat Stephen A. Douglas in the North and Southern Democrat John C. Breckinridge and Constitutional Unionist John Bell in the South. Although no candidate won 40% of the popular vote, in 81% of the counties some candidate won a majority of the vote. Furthermore, Abraham Lincoln would have won the electoral vote even if all opposition votes had gone to a single opponent. Combining information about the candidates' platforms with the state-by-state breakdown of the vote, we estimate one reasonable, if hypothetical, distribution of the 24 possible preference rankings by percentage as shown below. The percentage of voters ranking a given candidate first is known; it is simply the percentage of votes the candidate received in the election.

Percentage of Voters

	8.0	23.9	1.7	0.3	5.6	0.3	4.4	10.3	2.6	1.8	8.0	2.4
Lincoln	1	1	1	1	1	1	2	2	3	4	3	4
Douglas	2	2	3	4	3	4	1	1	1	1	1	1
Breckinridge	3	4	2	2	4	3	3	4	2	2	4	3
Bell	4	3	4	3	2	2	4	3	4	3	2	2

Percentage of Voters

	0.2	0.7	0.2	3.4	2.7	10.9	0.5	0.8	0.7	3.7	1.5	5.4
Lincoln	2	2	3	4	3	4	2	2	3	4	3	4
Douglas	3	4	2	2	4	3	3	4	2	2	4	3
Breckinridge	1	1	1	1	1	1	4	3	4	3	2	2
Bell	4	3	4	3	2	2	1	1	1	1	1	1

(a) Under our assumptions, who would have received the most votes in a runoff between Lincoln and Douglas?

(b) Who would be the Borda winner?

(c) Who, if anyone, is the Condorcet winner?

17. Which of the following choices would you expect to have single-peaked preferences among those voting?

(a) A review committee's merit ratings of an employee, where the choices are poor, good, excellent, and superior

(b) A restaurant chain's decision on how many restaurants to open in a region, where the choices are 1, 2, or 3 restaurants

(c) A group of people's preference rankings of the four vegetables corn, peas, green beans, and broccoli

18. Which of the following choices would you expect to have single-peaked preferences among those voting?

(a) A school board decision on student–teacher ratios, where the choices are 20, 25, 30, or 35 students per teacher

(b) A condominium committee's decision on what kind of trees to plant, where the choices are elm, maple, or pecan

(c) A group of student's preference rankings about the ideal day of the week to have an exam

19. If a candidate wins under plurality and is a Condorcet winner, must the candidate win a plurality election with runoff between the top two finishers, assuming there are no ties for the runoff positions? Give an example where this does not happen or explain why it always does.

20. In the voting method known as Nanson's method, the Borda counts of all candidates are computed. Next, those candidates with below average Borda counts (meaning a Borda count less than $v(c + 1)/2$, where v is the number of voters and c is the number of candidates) are eliminated. After the first round, the original preference rankings are used to construct a new set of preference rankings for the remaining candidates only. These preference rankings are then used to compute new Borda counts, those with below average counts are eliminated, and the process continues until a single candidate is left. Show that Nanson's method will select the Condorcet winner if there is one by relating the different head-to-head vote counts to the Borda count.

21. Suppose that preferences are single-peaked with respect to a given, known ordering of candidates along a line. Explain how to determine the Condorcet winner knowing only the first choices of the voters.

22. Construct an example of an election with three candidates that has a Condorcet winner with fewer first-place votes and fewer second-place votes than the Borda winner.

1.4 APPROVAL VOTING

Another voting method, called approval voting, dates back some two thousand years and was even used from 1294 to 1621 to select the Pope. In the 1970s, several political scientists and others, writing independently, reintroduced and analyzed approval voting, described as follows.

Approval Voting Method

With the **approval voting method**, voters indicate their approval or disapproval of each of the candidates. A ballot in an approval vote lists all the candidates' names, and voters check off all the candidates of whom they approve. The winner is declared to be the candidate with the highest approval count.

Voters can approve of from none to all of the candidates on a ballot, but to affect who is elected they should approve of at least one candidate and not approve of at least one candidate. Studies of approval voting indicate that voters tend to vote for at most half of all candidates. Approval voting is now used in the election of the Secretary General of the United Nations and the election of officers of some academic and professional societies.

The following example demonstrates how approval voting works.

EXAMPLE 1 *Approval Voting Method*

An 11-member committee has 5 members—Hart, Alvarez, Wolsey, Hanson, and Park—running for chair. The committee decides to use the approval voting method in the election. The results are listed in Table 1.13, where a √ indicates approval of the candidate.

TABLE 1.13 Approval Ballots in Committee Chair Election

| | *Number of Voters* | | | | | | |
	1	1	3	1	2	1	2
Hart			✓			✓	✓
Alvarez	✓		✓	✓		✓	✓
Wolsey					✓		✓
Hanson				✓			
Park		✓			✓	✓	✓

Which candidate is the winner of the election?

SOLUTION:

The approval vote count is Hart 6, Alvarez 8, Wolsey 4, Hanson 1, and Park 6. Therefore, Alvarez is the winner. ∎

We sometimes want to consider examples that compare the outcomes of elections using different voting methods. In order to incorporate approval voting in a table of preference rankings, we put checks next to the rankings of any candidates approved by the voters, as in the following example.

EXAMPLE 2 *Comparison of Voting Methods*

A city council needs to select a month in which to hold its first annual fair. The preference rankings of the council members along with the months of which they approve are shown in Table 1.14.

TABLE 1.14 Preference Rankings of Month for City Fair

	Number of Voters						
	3	1	1	2	1	2	1
June	1 ✓	1 ✓	2 ✓	3	3	2	2 ✓
July	2	3	1 ✓	1 ✓	1 ✓	3	3
August	3	2 ✓	3	2	2 ✓	1 ✓	1 ✓

(a) Which month would be selected if the approval method were used?
(b) Which month would win a plurality election?
(c) Which month is the Borda winner?

SOLUTION:

(a) In an approval election, June would receive 6 votes, July 4 votes, and August 5 votes, so June would win.
(b) The vote tallies in a plurality election would be 4 votes for June, 4 votes for July, and 3 votes for August, so a plurality election would result in a tie between June and July.
(c) June's Borda count is given by

$$4 \cdot 3 + 4 \cdot 2 + 3 \cdot 1 = 12 + 8 + 3 = 23.$$

Similarly, July's Borda count is 22 and August's Borda count is 21, so June is the Borda winner. ∎

Approval voting has entered the picture recently and has many supporters. It is particularly suitable when voters perceive two classes of candidates: those worthy of support and those not. Approval voting is less subject to manipulation than is Borda's method because each individual voter can only change the total vote count of a candidate by one point, whereas with Borda's method, the difference between first and last place can be several points. On the other hand, by grouping all candidates into one of two classes, those approved of and those not approved of, a voter may like one candidate much better than another and yet give them the exact same level of support.

In the 1980 presidential election, Ronald Reagan and Jimmy Carter faced independent candidate John Anderson. At one point in the campaign, Anderson's support was nearly 15%. However, by election day his support faded to 6.6%, as voters deserted him to choose between the two front-runners. In the election, Ronald Reagan won with 50.7% of the popular vote to Jimmy Carter's 41.0%. Thus, had the election been decided by popular vote, Ronald Reagan would have won with a majority in a plurality election. Whether he would have won under other voting methods is less clear. A *Time* magazine poll shortly before the election found that percentages finding each candidate "acceptable," presumably roughly equivalent to an affirmative approval vote, were Reagan 61%, Carter 57%, and Anderson 49%, much closer than the results of the popular vote. Exercises 13 and 14 at the end of this section ask you to build models consistent with these figures that will predict what the outcome of the election might have been had Condorcet's method or Borda's method been used.

In the following example, we look at another political situation to see how all the voting methods we have looked at play out.

EXAMPLE 3 *1912 Presidential Election*

In 1912, ex-President Theodore Roosevelt challenged incumbent President William Howard Taft for the Republican Party presidential nomination and lost. Roosevelt then ran for president under the Progressive Party or Bull Moose Party banner. The third major candidate in the 1912 presidential election was Democrat Woodrow Wilson. Also running was Socialist Eugene V. Debs, who made his best showing in his five attempts at the presidency. The results of the popular vote are given in Table 1.15.

TABLE 1.15 Popular Vote in 1912 Presidential Election

Wilson	Roosevelt	Taft	Debs	Other
6,293,152	4,119,207	3,486,333	900,369	241,902

We can roughly place the four candidates on a liberal–conservative spectrum as shown in Figure 1.3.

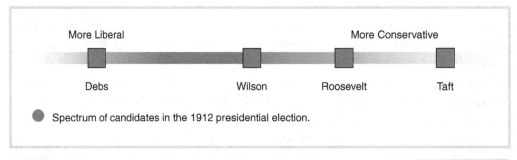

● Spectrum of candidates in the 1912 presidential election.

FIGURE 1.3 Spectrum of candidates in the 1912 presidential election.

We might assume that voters would be likely to vote for those candidates nearest to them on this spectrum. However, there were other pertinent factors. Taft was the incumbent. Former President Roosevelt, with a personality many people loved or hated, ran as an independent and entered under his specially formed Bull Moose Party. His positions on conservation and integration both attracted and alienated voters. Domestic issues were often separate from international policies such as isolationism; many voters focused on one or the other.

For simplicity, we consider only the votes for the top four candidates. Let us also assume that Debs supporters rank the candidates from first choice to last in this order: Debs, Wilson, Roosevelt, Taft. We further assume that the voters who did not vote for Debs would rank him last. Suppose also that 75% of Wilson supporters prefer Roosevelt to Taft, that 80% of Roosevelt supporters prefer Taft to Wilson, and that 85% of Taft supporters prefer Roosevelt to Wilson. Finally, we assume that all the voters approve of their first choices, 60% of the voters would also approve of their second choices, and no voter would approve of his or her third choice. Without accurate polling information, these are only plausible guesses. They are simplified, but still give us a qualitative idea of what *could* have happened. We want to answer the following questions about the 1912 presidential election:

(a) Who would have won a plurality of the popular vote had Roosevelt not run?
(b) Who would have won a plurality of the popular vote had Roosevelt won the Republican nomination, leaving Taft out of the election?
(c) Who would have won a majority of votes in a runoff between Wilson and Roosevelt?
(d) Who would have won by Borda's method?
(e) Would there be a Condorcet winner under the above assumptions?
(f) Who would have won an approval vote?

SOLUTION:

To answer all these questions, it is best to begin by determining the preference rankings of the voters. Based on our assumptions, the only possible preference rankings from first to last place are shown in Table 1.16.

TABLE 1.16 Possible Preference Rankings for the 1912 Presidential Election

Wilson	1	1	2	3	2	3	2
Roosevelt	2	3	1	1	3	2	3
Taft	3	2	3	2	1	1	4
Debs	4	4	4	4	4	4	1

Booker T. Washington dining with Theodore Roosevelt at the White House.

The 900,369 Debs supporters' preference rankings, based on our assumptions, were Debs first, Wilson second, Roosevelt third, and Taft last. Similarly, the number of voters ranking Wilson first, Roosevelt second, Taft third, and Debs last is

$$0.75 \cdot (6{,}293{,}152) = 4{,}719{,}864,$$

whereas the number of voters ranking Wilson first, Taft second, Roosevelt third, and Debs last is

$$6{,}293{,}152 - 4{,}719{,}864 = 1{,}573{,}288.$$

We compute estimates for the remaining four orderings similarly, and obtain the estimates shown in Table 1.17.

TABLE 1.17 Hypothetical Preference Rankings for the 1912 Presidential Election

	Number of Voters						
	4,719,864	1,573,288	823,841	3,295,366	522,950	2,963,383	900,369
Wilson	1	1	2	3	2	3	2
Roosevelt	2	3	1	1	3	2	3
Taft	3	2	3	2	1	1	4
Debs	4	4	4	4	4	4	1

Note that some of the entries in Table 1.17 have been rounded to give whole numbers. We can now answer our questions.

(a) Under our assumptions, had Roosevelt not run, Debs would have received 900,369 votes; Wilson would pick up 823,841 votes, for a total of 7,116,993 votes; and Taft would pick up 3,295,366 votes, for 6,781,699 votes. Therefore, Wilson would have won a plurality election.

(b) Dropping Taft from the election would yield 522,950 additional votes for Wilson, for a total of 6,816,102 votes, and 2,963,383 for Roosevelt, for a total of 7,082,590 votes. Debs would have received 900,369 votes. In this case, Roosevelt would have won a plurality election.

(c) Had Wilson and Roosevelt faced each other in a runoff, we could begin with the totals from part (b) and add Debs' votes to Wilson's total, giving Wilson a total of 7,716,471 votes, and making him the runoff winner.

(d) Wilson's Borda count would be

$$6{,}293{,}152 \cdot 4 + 2{,}247{,}160 \cdot 3 + 6{,}258{,}749 \cdot 2 = 44{,}431{,}586.$$

Similarly, Roosevelt's Borda count would be 45,519,783, Taft's Borda count would be 40,539,073, and Debs' Borda count would be 17,500,168. We see that Roosevelt is the Borda winner.

(e) Part (c) showed that Wilson would beat Roosevelt head-to-head. Clearly, Wilson would defeat Debs, who would be routed with only 900,369 votes. Finally, Wilson would defeat Taft, 8,017,362 votes to 6,781,699 votes. Thus, Wilson would be the Condorcet winner.

(f) Giving each candidate his votes plus 60% of his second-place votes, we find the approval vote totals to be Debs 900,369, Wilson 7,641,448, Roosevelt 8,729,155, and Taft 6,407,525. The approval vote winner would be Roosevelt. ∎

Given the simplified and speculative nature of our assumptions, only by trying a variety of equally reasonable assumptions can we get a good idea of the likely outcome of the 1912 presidential election under different methods of election. The more often such outcomes agree, the more likely they are to indicate what would have happened. In the exercises, we consider other sets of assumptions.

Approval voting has the advantages of being easy for voters to understand, being only moderately subject to strategic voting, and possibly leaving more voters satisfied with the outcome of an election. However, the approval voting method does not allow voters to specifically rank their preferences. Therefore, a candidate who is acceptable to, but not strongly liked by, many voters may win over a candidate who is exceptionally desirable to slightly fewer voters.

EXERCISES / FOR SECTION 1.4

1. A service club decided to use an approval vote in the election for president of the club. The five candidates for the position were McClain, Snyder, Freeman, Sanders, and Yang, and the results are given in the table.

Numbers of Voters

	1	1	1	2	2	1	1	1	1	1
McClain	✓			✓			✓			✓
Snyder		✓			✓	✓		✓	✓	
Freeman			✓		✓		✓	✓	✓	✓
Sanders							✓		✓	✓
Yang				✓		✓				✓

Who is the club's new president?

2. A professional society wants to hold its next annual convention in New Orleans, Las Vegas, San Francisco, or New York. The Executive Council, consisting of 12 members, decides to make the decision by approval ballot, and the results are listed in the table.

Number of Voters

	1	2	4	1	2	1	1
New Orleans	✓			✓	✓		✓
Las Vegas		✓		✓			✓
San Francisco					✓	✓	✓
New York			✓			✓	

Which city wins the vote?

3. The Institute of Electrical and Electronics Engineers (IEEE) used the approval method to elect its president in 1988. The results of this election are shown in the table, where the candidates' names have been replaced by the letters A, B, C, and D.

Number of Voters

	10,738	6,561	7,626	8,521	3,578	659	6,679	1,425	1,824	608	148	5,605	143	89	523
A	✓					✓	✓	✓			✓	✓	✓		✓
B		✓			✓			✓	✓		✓	✓		✓	✓
C			✓			✓		✓		✓	✓		✓	✓	✓
D				✓			✓		✓	✓		✓	✓	✓	✓

Source: Institute of Electrical and Electronics Engineers.

   (a) Find the winner of the election.
   (b) What percentage of the engineers voted in a way that could not help determine the winner regardless of how the others voted?

4. The members of a marching band vote to choose what time they want to hold an evening rehearsal. The choices are 6 P.M., 7 P.M., or 8 P.M. The results of an approval election are recorded here.

Number of Voters

	28	33	40	19	9	24
6 P.M.	✓			✓	✓	
7 P.M.		✓		✓		✓
8 P.M.				✓	✓	✓

Which time is the winner?

5. Three candidates are running for president of a union. A poll of the union membership indicates that if an approval election were held, the results would be as recorded here.

Percentage of Voters

	17	21	22	17	14	9
De Castro	✓			✓	✓	
Telger		✓		✓		✓
Segura			✓		✓	✓

Which candidate would win an approval election?

6. A toothpaste company decides to let the public determine a new flavor for children's toothpaste. The breakdown of the votes in an approval election are as follows:

Percentage of Voters

	9	15	10	8	9	4	6	5	3	1	7	4	9	2	8
Grape	✓				✓	✓	✓				✓	✓	✓		✓
Strawberry		✓			✓			✓	✓		✓	✓		✓	✓
Cherry			✓			✓		✓			✓	✓		✓	✓
Orange				✓		✓		✓	✓		✓	✓	✓	✓	

Which flavor is chosen?

7. The preference rankings of a class of 16 college students for the minimum voting age and the ages of which they approve are shown in the table.

Number of Voters

	1	1	1	1	1	3	2	2	3	1
Sixteen	4	6	5	5	6	6	6	6	6	6
Seventeen	2 ✓	2 ✓	3	3	3 ✓	5	5	5	5	5
Eighteen	1 ✓	1 ✓	1 ✓	1 ✓	1 ✓	1 ✓	1 ✓	1 ✓	4	4
Nineteen	3 ✓	3	2	2 ✓	2 ✓	2	2 ✓	2 ✓	3	3
Twenty	6	4	4	4	4	3	3	3 ✓	2	2 ✓
Twenty-one	5	5	6	6	5	4	4	4	1 ✓	1 ✓

(a) Which age would win a plurality vote?

(b) Which age would win a plurality vote with runoff between the top two finishers?

(c) Which age would win the Borda count?

(d) Which age, if any, is the Condorcet winner?

(e) Which age would win an approval vote?

8. The preference rankings of a class of 17 college students for the minimum legal drinking age and the ages of which they approve are shown in the table.

Number of Voters

	1	1	2	2	1	1	1	6	1	1
Sixteen	1 ✓	3	6	6	6	6	6	6	6	6
Seventeen	2	2	5	5	5	5	5	5	5	5
Eighteen	3	1 ✓	1 ✓	1 ✓	1 ✓	4	4	4	4	4
Nineteen	4	4	2	2 ✓	2 ✓	3	3 ✓	3	3	3 ✓
Twenty	5	5	3	3 ✓	3 ✓	1 ✓	1 ✓	2	2 ✓	2 ✓
Twenty-one	6	6	4	4	4 ✓	2 ✓	2 ✓	1 ✓	1 ✓	1 ✓

(a) Which age would win a plurality vote?

(b) Which age would win a plurality vote with runoff between the first- and second-place finishers?

(c) Which age would win the Borda count?

(d) Which age, if any, is the Condorcet winner?

(e) Which age would win an approval vote?

9. The partners of a law firm must decide which one of five junior members to promote to partner. The preference rankings of the partners and the candidates of which they approve are included in the table.

Number of Voters

	2	1	1	1	1	3	1	1
Douglas	4	4	5	4	5	4	5	5
Holmes	1 ✓	1 ✓	1 ✓	2	2 ✓	2	2 ✓	3
Cochrane	5	5	4	3	4 ✓	5	4	4
Mason	3	3	2 ✓	1 ✓	1 ✓	3	3 ✓	2 ✓
Bailey	2	2 ✓	3 ✓	5	3 ✓	1 ✓	1 ✓	1 ✓

(a) Which member would win a plurality vote?

(b) Which member would win a plurality vote with runoff between the first- and second-place finishers?

(c) Which member would win the Borda count?

(d) Which member, if any, is the Condorcet winner?

(e) Which member would win an approval vote?

(f) Could the voter whose first choice was Holmes and second choice was Mason have obtained a preferable result in an approval vote by voting strategically if the others voted as shown in the table?

10. Due to prison overcrowding, a parole board must release one of four minor offenders before he has served his full sentence. The board votes to determine which prisoner will be released. The preference rankings of the parole board members and the candidates of which they approve are included in the table.

Number of Voters

	1	1	1	1	1	1	1
Ford	1 ✓	2 ✓	4	3	2	3	3
Robbins	3	1 ✓	3 ✓	4	3	2	2 ✓
Cage	2 ✓	3	1 ✓	1 ✓	4	4	4
Bacon	4	4	2 ✓	2 ✓	1 ✓	1 ✓	1 ✓

(a) Which prisoner would win a plurality vote?
(b) Which prisoner would win a plurality vote with a runoff between the top two finishers?
(c) Which prisoner would win the Borda count?
(d) Which prisoner, if any, is the Condorcet winner?
(e) Which prisoner would win an approval vote?
(f) Could the voter whose first choice was Robbins have obtained a preferable result in an approval vote by voting strategically if the others voted as shown in the table?

11. A condominium association must vote to determine what color to paint the wood trim of a building. The three choices are white, green, and ivory. The preference rankings of the association members and the colors of which they approve are shown in the table.

Number of Voters

	4	3	2	1	2	2	4	2	1
White	1 ✓	1 ✓	1 ✓	2 ✓	3	3	2	3	3
Green	2	3	3	1 ✓	1 ✓	1 ✓	3	2	2 ✓
Ivory	3	2	2 ✓	3	2	2 ✓	1 ✓	1 ✓	1 ✓

(a) Which color would win a plurality vote?
(b) Which color would win a plurality vote with runoff between the top two finishers?
(c) Which color would win the Borda count?
(d) Which color, if any, is the Condorcet winner?
(e) Which color would win an approval vote?
(f) Could the five voters with first choice white and second choice ivory have obtained a preferable result in an approval vote by voting strategically if the others voted as shown in the table?

12. A PTA board must determine what kind of fund-raiser it will hold. The choices have been narrowed down to holding a carnival, selling magazines, selling

wrapping paper, or holding an auction. The preference rankings of the 11 board members and the choices of which they approve are shown here.

Number of Voters

	1	2	1	1	1	1	3	1
Holding a carnival	1 ✓	1 ✓	3	4	3	4	2 ✓	2
Selling magazines	4	4	1 ✓	1 ✓	2 ✓	2 ✓	3	3
Selling wrapping paper	3	3	2 ✓	2 ✓	1 ✓	1 ✓	4	4
Holding an auction	2	2 ✓	4	3	4	3 ✓	1 ✓	1 ✓

(a) Which fund-raiser would win a plurality vote?
(b) Which fund-raiser would win a plurality vote with runoff between the top two choices?
(c) Which fund-raiser would win the Borda count?
(d) Which fund-raiser, if any, is the Condorcet winner?
(e) Which fund-raiser would win an approval vote?
(f) Could the three board members whose first choice was holding a carnival have obtained a preferable result in an approval vote by voting strategically if the other members voted as shown in the table?

In Exercises 13-16, we consider outcomes of hypothetical approval votes in Connecticut's 2006 Senatorial election. Having been defeated by Ned Lamont in the Democratic primary, Joseph Lieberman ran and won as an independent. The Republican candidate was Alan Schlesinger. From CNN exit polls of Democratic, Republican, and Independent voters, we hypothesize the following preference rankings.

Percentage of Voters

	33	18	36	4	7	2
Joseph Lieberman	1	1	2	3	2	3
Ned Lamont	2	3	1	1	3	2
Alan Schlesinger	3	2	3	2	1	1

13. Who wins an approval vote if everyone approves of only his or her first choice?
14. Who wins an approval vote if everyone approves of both his or her first and second choices?
15. Who wins an approval vote if everyone approves of his or her first choice and 60% of those with any ranking approve of their second choice?
16. Who wins an approval vote if everyone approves of his or her first choice and 30% of those with any ranking approve of their second choice?

17. Counting only votes cast for the top three candidates in the 1980 presidential election, Ronald Reagan received 51.6% of these votes, Jimmy Carter received 41.7%, and John Anderson received 6.7%. Assume that these percentages represent the first choices of the voters (even though it is clear that some Anderson

supporters voted for either Reagan or Carter in order to have more potential effect on the election). Assume that their approval counts were Reagan 61%, Carter 57%, and Anderson 49%, and that every voter approved of either one or two candidates.

(a)   Explain why Reagan must be the Condorcet winner.

(b)   What percentage of people had Ronald Reagan as their second choice *and* approved of him?

(c)   What percentage of people had Jimmy Carter as their second choice *and* approved of him?

(d)   What percentage of people had John Anderson as their second choice *and* approved of him?

(e)   Estimate the percentage, rounded to the nearest whole number percentage, of second-place votes each candidate would get from among *all* the voters. To do this use your answers to parts (b) through (d) and assume that the percentage of second-place votes a candidate would get among all the voters is equal to the percentage of second-place votes that candidate got from among those voters who approved of their second choice. (Be sure your percentages add up to 100%.)

(f)   Estimate the results of an election by Borda's method.

18.   John Anderson's support in the 1980 presidential election was as high as 15% in the preelection polls, before fading to 6.6% in the actual election. We will ignore all candidates other than Anderson, Ronald Reagan, and Jimmy Carter. Assume that the drop-off in the election represented strategic voters who still ranked Anderson first (rather than any disillusionment with Anderson). Assume that the remaining first-place ranks were proportional to the actual vote total, so that Reagan was ranked first by 47% and Carter by 38%. Assume that their approval counts were Reagan 61%, Carter 57%, and Anderson 49%, and that every voter approved of either one or two candidates.

(a)   What percentage of people had Ronald Reagan as their second choice *and* approved of him?

(b)   What percentage of people had Jimmy Carter as their second choice *and* approved of him?

(c)   What percentage of people had John Anderson as their second choice *and* approved of him?

(d)   Estimate the percentage, rounded to the nearest whole number percentage, of second-place votes each candidate would get from among *all* of the voters. To do this use your answers to parts (a) through (c) and assume that the percentage of second-place votes a candidate would get among all the voters is equal to the percentage of second-place votes that candidate got from among those voters who approved of their second choice. (Be sure your percentages add up to 100%.)

(e)   Estimate the results of an election by Borda's method.

(f)   What fraction of those voters ranking Anderson first would need to rank Reagan second in order for Reagan to defeat Carter in a head-to-head comparison?

(g) What fraction of those voters ranking Carter first would need to rank Reagan second in order for Reagan to defeat Anderson in a head-to-head comparison?

(h) In light of your answers to parts (d), (f), and (g), does it seem likely that Reagan would be a Condorcet winner?

In Exercises 19–21, we consider the 1912 presidential election as we did in Example 3. The popular vote count for the top four candidates is shown here.

Wilson	Roosevelt	Taft	Debs
6,293,152	4,119,207	3,486,333	900,369

Again, we just look at these top four candidates. For brevity we use the notation DWRT to denote a preference ranking of Debs first, Wilson second, Roosevelt third, and Taft last, and denote the other possible preference rankings similarly. For the given situation answer the following.

(a) Who would have won a plurality of the popular vote had Roosevelt not run?

(b) Who would have won a plurality of the popular vote had Roosevelt won the Republican nomination, leaving Taft out of the election?

(c) Who would have won the majority of votes in a runoff between Wilson and Roosevelt?

(d) Who would have won by Borda's method?

(e) Which candidate, if any, is a Condorcet winner under the assumptions?

(f) Who would have won an approval vote?

(We remark that we use "his" in the restrictive sense at the end of questions 19 and 20 because women were not allowed to vote in Presidential elections until 1920.)

19. Assume that all Debs voters ranked the candidates DWRT. Also assume that 10% of Wilson voters ranked the candidates WDRT, 20% ranked them WTRD, and the rest of the Wilson voters ranked them WRTD. Assume that 75% of Roosevelt voters ranked them RTWD and the rest RWTD, and that 80% of Taft voters ranked them TRWD and the rest TWRD. Finally, assume that all voters approve of their first choices, 25% of the voters approve of their second choices, and no voter approves of his third or fourth choices.

20. Assume that all Debs voters ranked the candidates DWRT and that the voters who did not vote for Debs would rank him last. Assume that 45% of Wilson supporters ranked the candidates WRTD, that 70% of Roosevelt supporters ranked the candidates RTWD, and that 70% of Taft supporters had the ranking TRWD. Finally, assume that all voters approve of their first choices, 50% of the voters approve of their second choices, and that no voter approves of his third or fourth choices.

21. Assume that all Debs voters ranked the candidates DWRT. Assume that 20% of Wilson voters ranked the candidates WDRT and the rest of the Wilson voters ranked them WRTD, that 85% of Roosevelt voters ranked them RTWD and

the rest RWTD, and that 90% of Taft voters ranked them TRWD and the rest TWRD. Finally, suppose that 75% of voters in each group would support their second choice in an approval vote and that none would support their third or fourth choices.

In Exercises 22–25, we look at state elections. The Progressive Party nominated candidates for many offices in 1912. We consider possible outcomes under different voting methods of several governors' races that had moderately close races between the Democratic, Republican, and Progressive Party candidates in 1912. We will ignore other candidates in these races. The votes are given either as raw numbers or as percentages. Assume that 60% of Democratic voters ranked the Progressive candidate second, that 70% of Republican voters ranked the Progressive candidate second, that 65% of Progressive candidates ranked the Republican candidate second, and that 40% of those with a given ranking would vote to approve of only their first and second choice, whereas the remaining 60% would vote to approve of only their first choice in an approval vote. Determine

    (a)   the winner of a runoff election,

    (b)   the Borda winner,

    (c)   the Condorcet winner, if there is one,

    (d)   the winner of an approval vote.

22. Illinois: E. Dunne (Democrat) 443,120, C. Deneen (Republican) 318,469, F. Funk (Progressive) 303,401

23. Connecticut: S. Baldwin (Democrat) 78,264, J. Studley (Republican) 67,531, H. Smith (Progressive) 31,020

24. Utah: J. Tolten (Democrat) 35.3%, W. Spry (Republican) 41.6%, N. Morris (Progressive) 23.1%

25. Michigan: W. Ferris (Democrat) 37.5%, A. Musselman (Republican) 32.9%, L. Watkins (Progressive) 29.6%

In Exercises 26 and 27, we consider the 1970 New York Senate election, which was decided by the plurality method among three candidates. James Buckley (Conservative) with 39% of the vote defeated Richard Ottinger (Democrat) with 37% of the vote and Charles Goodell (Republican and Liberal Parties) with 24% of the vote. (The Liberal Party consisted mostly of more liberal Republicans, so it would land between the Democrats and the Conservatives on the liberal–conservative spectrum.) Assume that preferences are single-peaked, with Goodell in the middle.

26. Suppose that among Goodell supporters, 60% have Ottinger as a second choice and 40% have Buckley as a second choice. Suppose also that 40% of voters with any particular preference ranking approve of their second choice, that all approve of their first choice, and that none approve of their third choice.

    (a)   Who would win a runoff between Buckley and Ottinger?

    (b)   Who is the Condorcet winner?

    (c)   Who wins an approval vote?

27. Suppose that Goodell supporters split 50%-50% for their second choice and that, for each ranking, half of the voters with this ranking approve of their

second choice, all approve of their first choice, and none approve of their third choice.

(a)   Who would win a runoff between Buckley and Ottinger?

(b)   Who is the Condorcet winner?

(c)   Who wins an approval vote?

28. The 1980 New York Senate election, decided by plurality, saw Republican Alphonse D'Amato win with 45% of the vote, trailed by Democrat Elizabeth Holtzman with 44%, and Jacob Javits, running under the Liberal Party banner after losing the Republican nomination, with 11%. Javits was in the middle of the liberal–conservative spectrum. However, concerns about his age and health probably prevented preferences from being single-peaked. An ABC News poll estimated the results of a potential approval vote to be Holtzman 60%, D'Amato 56%, and Javits 49%. Assume that voters always approved of their first choice and never approved of their third choice.

(a)   For each candidate, what percentage of the voters placed that candidate second *and* approved of him or her?

(b)   What is the smallest possible percentage of voters who departed from rankings based on the liberal–conservative spectrum?

(c)   Estimate the percentage of second-place votes each candidate would get from among *all* the voters. To do this use your answer to part (a) and assume that the percentage of second-place votes a candidate would get among all the voters is equal to the percentage of second-place votes the candidate got from among those voters who approved of their second choice. (Be sure your percentages add up to 100%.)

(d)   Estimate the results of an election by Borda's method.

29. Show that you, as a single voter, make the most head-to-head distinctions between pairs of candidates (approving of one candidate and disapproving of the other) when you approve of as close to half of all candidates as possible.

1.5   THE SEARCH FOR AN IDEAL VOTING SYSTEM

Each voting method discussed in this chapter can yield outcomes that seem less than ideal. For example, if the plurality method is used with a slate of more than two candidates, then the winner may well be a candidate that the majority of the voters find highly undesirable. We have also seen that a Condorcet winner may not win an election under Borda's method and may not even make the runoff in a plurality election with a runoff. On the other hand, there may not be a Condorcet winner. It seems reasonable to ask for a better voting system, one that always satisfies certain desirable properties. In this section, we follow the ideas of Nobel Laureate Kenneth Arrow (1921–), who, beginning in the 1940s, looked at the general properties of ways of ordering choices among public policies. His theory applies equally well to elections or to any issues of group choice. The results of his work depend only on the general properties that are assumed and are independent of the actual voting scheme used to tally the results.

A **voting system** takes a collection of ballots and combines them according to the rules of the method to produce a winner (or winners). The plurality, plurality with runoff, and Borda methods are all voting systems that can be applied to ballots where candidates are ranked. Ballots for the approval method consist of those names of which the voter approves, so ballots for this voting system are not preference rankings, but the approval method is still a voting system. Arrow set out a list of four properties that someone might wish an ideal voting system to satisfy. He considers only ballots that are preference rankings with ties in the rankings permitted. He further assumes that voters are consistent—they either always rank A higher than B, always rank A lower than B, or always rank A and B the same, regardless of what other candidates are in an election.

The first property listed by Arrow is called **universal domain**, and it stipulates that any ordering of the candidates is allowed. In other words, there are no restrictions placed on the rankings of the candidates a voter may choose.

The second property, called **Pareto optimality**, requires that if all the voters prefer candidate A to candidate B, then the group choice should not prefer B to A. This is a minimal condition for the choice to reflect the will of the constituents rather than be imposed by some outside entity.

Along the same lines, a third property is that no one individual voter's preferences totally determine the group choice. This condition is referred to as **nondictatorship**.

The fourth and final property in Arrow's system, called **independence from irrelevant alternatives**, requires that if a group of voters chooses candidate A over candidate B, then the addition or subtraction of other choices or candidates should not change the group choice to B. For instance, if the group would select apple pie when the choices are apple pie and pumpkin pie, it should not select pumpkin pie when the choices are expanded to apple pie, pumpkin pie, and pecan pie. However, it would be perfectly reasonable for the group to select pecan pie over both apple pie and pumpkin pie. Of the four properties, this last one is by far the most debatable. In its favor, a choice between A and B should not depend on what other choices happen to be available—other choices that may be arbitrary or deliberately chosen to achieve a particular outcome. On the other hand, it is only by comparison with other possibilities that voters' perceptions of differences between candidates can be brought to light.

Arrow's acclaimed theorem is that no voting method based on rankings can satisfy these four properties. More precisely, it can be stated as follows.

Arrow's Impossibility Theorem

There is no voting method based on rankings that satisfies the properties of universal domain, Pareto optimality, nondictatorship, and independence from irrelevant alternatives.

So if we consider the four properties to be necessary for an ideal voting method, then no ideal voting method exists.

Arrow and others have made changes in the required properties and have also relaxed the condition that ballots be rankings. These four properties are only some of many possibilities. Another possible axiom is **anonymity**, which requires that each voter's ballot receives the same treatment or weight. Thus, interchanging the ballots of two or more voters would not change the outcome of the election. This egalitarian assumption seems a most reasonable one for a democracy. Another, **neutrality**, requires that each candidate receive equal treatment. Thus, interchanging all voters' rankings of two candidates would also exchange the election outcome for the two candidates. For example, favoritism for a particular spot on the ballot is not permitted. All the voting methods we have discussed are both anonymous and neutral.

We have also seen in the text and exercises that voters can sometimes obtain better results by strategic voting—in other words, by casting ballots that do not reflect their true rankings. In the early 1970s, Allan Gibbard and Mark Satterthwaite independently proved that in any voting system, other than a dictatorship, that is based on rankings, there will be instances where one or more voters may obtain a more desirable outcome through strategic voting. One way of proving their theorem is to show that a system in which no one can gain through strategic voting must satisfy all the properties of Arrow's theorem—conditions that cannot simultaneously hold. Determining a voter's best decisions, allowing for strategic voting, falls into the study of game theory.

EXAMPLE 1 *Plurality Method and Pareto Optimality*

Explain why the plurality method satisfies the Pareto optimality property.

SOLUTION:

If every voter prefers candidate A to candidate B, then B cannot get any first-place votes, so cannot be the plurality winner.   ∎

EXAMPLE 2 *Plurality Method and Independence from Irrelevant Alternatives*

Give an example that shows how the plurality method can violate the independence from irrelevant alternatives property.

SOLUTION:

Suppose A defeats B in a plurality election by 4 to 3. Now suppose a new candidate C enters the picture and that two of A's supporters rank C first, whereas the other voters all rank C last. The new plurality vote would be B 3, A 2, and C 2, so the group of voters now favors B over A, and B wins the election.

The rankings of the seven voters are shown in Table 1.18. Note that it is consistent with the original 4 to 3 tally for A over B.

TABLE 1.18 Preference Rankings for Plurality with C Added

	Number of Voters		
	2	2	3
A	2	1	2
B	3	2	1
C	1	3	3

Although the solution to Example 2 is short and easy to understand, it is deceptively simple because it may take quite a bit of effort to come up with such an example. Solving these kinds of problems usually requires a bit of trial-and-error, using what you learn from any unsuccessful tries. It often pays to look for examples with small numbers. In the case of searching for a solution to Example 2, we started with A barely defeating B so that it would not take much to reverse the result. In fact, if A defeats B by a single vote, A need only lose two votes to a third candidate C in order for B to get more votes than A. If A initially defeated B by 2 to 1, then C would end up winning the election. If A initially defeated B by 3 to 2, then B and C end up tied for the win. Therefore, we moved on to try the example of A defeating B by 4 to 3, which worked.

After learning of impossibility theorems such as those of Arrow and of Gibbard and Satterthwaite, we might be tempted to abandon the search for new and improved voting systems. However, a great deal of current research attempts to find voting methods that satisfy slightly weaker conditions or that satisfy a set of conditions a large proportion of the time.

In practice, the voting methods we have studied work well most of the time. In many elections, almost any voting method one might care to use will result in the same winner. Unfortunately, the plurality method, which is the one most commonly used in the United States, seems to result in controversial outcomes more often than other methods.

E X E R C I S E S FOR SECTION 1.5

1. Explain why the method of selecting the candidate with the fewest last place votes satisfies the Pareto optimality property.

2. Explain why Borda's method satisfies the Pareto optimality property.

3. Give an example that shows how Borda's method can violate the independence from irrelevant alternatives property.

4. Give an example that shows how the plurality method with runoff between the top two finishers can violate the independence from irrelevant alternatives property.

5. Can an anonymous system (with more than one voter) be a dictatorship? If so, give an example; if not, explain why not.

6. When either the U.S. House of Representatives or the Senate considers a bill, amendments are voted up or down one at a time and after consideration of all amendments, the final bill, amended or not, is voted on.
   (a) Explain why this method is anonymous.
   (b) Give an example to show that this method is not neutral. (Here, neutrality would mean that the final result does not depend on the order of consideration.)
   (c) If a particular bill must pass both House and Senate, show that when one takes into account all representatives and all senators, the process is not anonymous.

7. In approving presidential nominations, the Senate votes yes or no on a particular nomination, the decision being by majority. If the nominee is approved, the process ends. Otherwise, it moves on to a new nomination. In some sense, this is similar to an approval ballot if one assumes that a senator would vote for any candidate of whom he or she approves.
   (a) Is this method anonymous?
   (b) Is this method neutral? (Here, we assume that the order in which nominees are presented, until one is confirmed, is fixed.)

8. Suppose that a deliberative body (with more than one member) only votes on certain monetary matters for which the preferences are single-peaked and that the winning choice is the Condorcet winner. Clearly the property of universal domain is violated.
   (a) Explain why the system satisfies the Pareto optimality property.
   (b) Explain why the system satisfies the independence from irrelevant alternatives property.

9. Consider the election method that chooses the first name on the list when the candidates are listed alphabetically.
   (a) Does this method satisfy universal domain?
   (b) Does this method satisfy non-dictatorship?
   (c) Does this method satisfy Pareto optimality?
   (d) Does this method satisfy independence from irrelevant alternatives?
   (e) Does this method satisfy anonymity?
   (f) Does this method satisfy neutrality?

10. Consider an election method where we drop all candidates with more last place votes than first place votes, re-compute first and last place votes for all remaining candidates as in a runoff election, drop all candidates with more last place votes than first place votes, . . . , until there is a winner.
    (a) Does this method satisfy universal domain?
    (b) Does this method satisfy non-dictatorship?

(c) Does this method satisfy Pareto optimality?

(d) Does this method satisfy independence from irrelevant alternatives?

(e) Does this method satisfy anonymity?

(f) Does this method satisfy neutrality?

11. Consider a sequential election method, where the first two candidates on the ballot face off in a head-to-head election, the winner facing the third candidate on the ballot in a head-to-head election, and so forth.

(a) Does this method satisfy universal domain?

(b) Does this method satisfy non-dictatorship?

(c) Does this method satisfy Pareto optimality?

(d) Does this method satisfy independence from irrelevant alternatives?

(e) Does this method satisfy anonymity?

(f) Does this method satisfy neutrality?

12. Consider an election method that drops candidates listed at the bottom of the ballot, if necessary and one at a time, until there is a Condorcet winner, who is the winner of the election.

(a) Does this method satisfy universal domain?

(b) Does this method satisfy non-dictatorship?

(c) Does this method satisfy Pareto optimality?

(d) Does this method satisfy independence from irrelevant alternatives?

(e) Does this method satisfy anonymity?

(f) Does this method satisfy neutrality?

Exercises 13–16 examine another desirable property of a voting method. A voting method or system is called *monotone* if a candidate never does worse when one or more voters raise the candidate in their rankings while all other candidates maintain the same positions relative to one another.

13. Explain why the plurality method is monotone.

14. Explain why Borda's method is monotone.

15. In Example 3 of Section 1.3, we looked at the 1990 Louisiana election for Governor. The method of election was plurality with runoff. We hypothesized the following preference ranking.

	Percentage of Voters					
	9	28	11	23	23	6
Edwards	1	1	2	3	2	3
Duke	2	3	1	1	3	2
Roemer	3	2	3	2	1	1

(a) Show that if the 11% that ranked the candidates Duke, Edwards, Roemer instead ranked them Edwards, Duke, Roemer, then Edwards would have lost the election, violating the monotonicity property.

(b) What is the smallest percentage that could change their ranks from Duke, Edwards, Roemer to Edwards, Duke, Roemer and cause Edwards to lose?

16. The 2006 Peruvian Presidential Election was determined by plurality with runoff. The winner was Alan García Pérez, who finished second in the initial round of voting. Several pre-election polls indicated that the third place candidate, Lourdes Flores Nano, would have been the Condorcet winner. Ignoring 13 candidates who collectively received 3.7% of the vote, the results were as follows.

	Ollanta Humala Tasso	Alan García Pérez	Lourdes Flores Nano	Martha Chavez Cossio	Valentín Paniagua Corazao	Humberto Lay Sun
Percentage of Voters in 1st Round	31.8	25.3	24.7	7.7	6	4.5
Percentage of Voters in Runoff	45.3	54.7				

Construct a set of preference rankings for these six candidates that leads to the percentages above, that has Flores Nano as the Condorect winner, and such that if some voters had raised García Pérez in their rankings then García Pérez could have lost.

Exercises 17-23. Although approval voting is not based on rankings, it is possible to consider whether it satisfies the properties we have discussed. Assume that each voter has an internal ranking of the candidates and will never disapprove of a candidate ranked above a candidate of which he or she approves.

17. (a) How should universal domain be defined in the context of approval voting?
    (b) Does approval voting satisfy universal domain?
18. Does approval voting satisfy Pareto optimality?
19. Does approval voting satisfy independence from irrelevant alternatives?
20. Does approval voting satisfy non-dictatorship?
21. Does approval voting satisfy anonymity?
22. Does approval voting satisfy neutrality?
23. (a) How should one interpret the term *monotone* (see Exercises 13-16) mean in the context of approval voting?
    (b) Is approval voting monotone?

WRITING EXERCISES

1. Suppose an election involves three candidates: one liberal, one moderate, and one conservative. Polls indicate that the liberal has 41% support, the moderate 20%, and the conservative 39%. Explain what you would expect to happen in an election decided by each of the voting methods we studied. Which method do you feel would yield the fairest result in this situation? Support your answer with an explanation.

2. A plurality election followed by a runoff can be conducted by having the voters fill out a preference ballot. However, runoff elections typically are held by having the voters fill out separate ballots for each stage of the process, and there may be several days or weeks between the separate ballots. Discuss the pros and cons of these two options.

3. By asking voters to support all the candidates acceptable to them, approval voting attempts to shift the voter's view from thinking of the glass as half empty to thinking of it positively as half full. Supporters of the approval voting method claim it will reduce negative campaigning, increase voter turnout, and increase the viability of minority candidates. In your view, how likely are these claims to be true?

4. Of the voting methods of plurality, plurality with a runoff between the top two candidates, Borda's method, or the approval method, which do you feel is the best method overall for an election with more than two candidates? Which method do you feel is the weakest? Support your answers with full explanations.

5. All voting methods based on preference rankings violate one of the four properties of Arrow's Impossibility Theorem. Which property would you drop as a requirement of a voting system and why?

6. On the front page of the August 16, 1995 *Los Angeles Times*, writer K. C. Cole wrote "Mathematicians do not agree on the best system. But they have no problem pointing their fingers at the worst: the plurality systems used in most U.S. elections." On page 316 of his book, *The Next American Nation*, author Michael Lind writes, "Our archaic first-past-the-post plurality electoral system preserves a two-party monopoly rejected by a growing number of alienated American voters, forces many Americans to waste their votes and effectively disenfranchises substantial numerical minorities. There are practically no arguments in its favor, other than the fact that it is more than two hundred years old." Comment on these two quotes.

7. Suppose you asked a group of people to give their preference rankings on the ideal time to begin work, where the choices are 6, 7, 8, 9, or 10 A.M. Would you expect the set of preference rankings to be single-peaked? Support your answer with an explanation.

PROJECTS

1. When there are three candidates in an election, there is usually a Condorcet winner. With more candidates, the likelihood of a Condorcet winner decreases, but even with five candidates there is more likely than not a Condorcet winner.

   (a) Think of five issues with three "candidates" for which elections could be held in which you might *not* expect a Condorcet winner. Ask 15 people to fill out preference ballots for each of the five elections to see how often you get a Condorcet winner.

   (b) Repeat the process in part (a) for issues with four candidates.

   (c) Repeat the process in part (a) for issues with five candidates.

   (d) Are the results as you expected? Based on your limited results, what effect did increasing the number of candidates seem to have on the likelihood of a Condorcet winner?

2. Different voting methods sometimes yield the same winner and sometimes yield different winners. To study how they differ in practice, think of 10 issues with four "candidates" for which elections could be held. Ask 15 people to fill out preference ballots and indicate how they would vote on an approval ballot for each of the 10 elections. For each election, find

   (a) the winner of a plurality vote,

   (b) the winner of a plurality vote with runoff between the top two finishers,

   (c) the Borda winner,

   (d) the Condorcet winner, if there is one,

   (e) the winner of an approval vote.

   Discuss the results of the different voting methods for your elections.

3. In 1861, Thomas Hare introduced a voting method now known as the Hare method or the single transferable vote method. This method is simply a plurality with runoff method in which in each stage the candidate (or candidates, if a tie) with the fewest votes is eliminated, and a runoff is held between the remaining candidates.

   (a) Construct examples of elections showing that the Hare method does not satisfy independence from irrelevant alternatives, is not monotone, and may not select the Condorcet winner, if there is one. (For the definition of monotone, see Exercises 10–12 in Section 1.5.)

   (b) Find some examples of elections in which the Hare method is actually used.

4. The 1968 election was the last in which a third-party candidate actually carried a state. Running against Republican Richard Nixon and Democrat Hubert Humphrey, George Wallace, running under the American Independent party label after seeking the Democratic nomination, won five southern states. After dropping the one-third of one percent who voted for some other candidate, the actual vote is given in the first column of the table. A Gallup

poll estimated the breakdown by political party as shown in the final three columns of the table.

	Percentage of All Votes	Percentage of Republican Votes	Percentage of Democratic Votes	Percentage of Independent Votes
Nixon	43.5	86	12	44
Humphrey	42.9	9	74	31
Wallace	13.6	5	14	25

Source: Gallup Opinion Index.

Although Nixon won handily in the Electoral College, the popular vote was razor close.

(a) To analyze the election results, we first find the percentage of voters who are Republican, Democrat, and Independent. Letting R, D, and I denote the percentage of Republican, Democratic, and Independent voters, respectively, we can estimate these percentages by solving the set of equations

$$0.86R + 0.12D + 0.44I \approx 43.5,$$
$$0.09R + 0.74D + 0.31I \approx 42.9,$$
$$0.05R + 0.14D + 0.25I \approx 13.6,$$
$$R + D + I = 100.$$

Explain why R, D, and I should satisfy these four equations.

(b) Show that an approximate solution to the set of equations in part (a) is given by $R \approx 32.6\%$, $D \approx 44.4\%$, and $I \approx 23.0\%$.

(c) Suppose that within each of the three classes of voters, the second choices are proportional to the percentages supporting the other two candidates. For instance, of the Republicans voting for Nixon, $9/(9 + 5)$ would have Humphrey as their second choice. Create a table showing the percentage of voters having each possible preference ranking. Estimate who would have won the popular vote in a runoff between Nixon and Humphrey, and estimate who would have won the popular vote in an election between Nixon and Wallace.

5. Create a spreadsheet or computer program that will calculate the results of an election for one or more of the methods we have studied.

6. In 1956, the House of Representatives considered allocating funds to the states for school construction. New York Democrat Adam Clayton Powell introduced an amendment to withhold funds from any state not complying with Supreme Court decisions, which was particularly aimed at southern states maintaining segregated schools. The Powell amendment passed; the amended bill was defeated. The votes of those representatives voting on both the amendment and

the funding bill broke down as follows, with the first vote on the amendment (Y for yes, N for no).

	YY	YN	NY	NN
Republicans	52	96	23	22
Southern Democrats	0	0	3	101
"Northern" Democrats	77	0	39	3

Source: Congressional Record.

There are six possible rankings of the alternatives of the original allocation, the amended bill, and no bill.
(a) Which rankings are possible for those who voted yes on both including the amendment and the amended bill?
(b) Which rankings are possible for those who voted yes on including the amendment and no on the amended bill?
(c) Which rankings are possible for those who voted no on including the amendment and yes on the amended bill?
(d) Which rankings are possible for those who voted no on both including the amendment and the amended bill?

It seems reasonable to assume that no representative would vote for the bill only with the Powell amendment. Further assume, perhaps less reasonably, that when preferences between the original bill and no bill remain unclear, Democrats favor the bill and Republicans oppose it.
(e) Which of the three alternatives wins a plurality?
(f) Which choice would win a runoff election between the top two alternatives?
(g) Which option is the Borda winner?
(h) Is there a Condorcet winner?
(i) Could any group with a given ranking have achieved a preferable result by voting strategically if the others voted as shown in the table?

It was suggested by President Truman and others in the Democratic leadership that the Republicans who voted first for the Powell amendment and then against the entire bill were voting strategically on the amendment in order to defeat the bill and also to embarrass the Democrats. The facts are ambiguous and political scientists do not agree whether there was significant strategic voting. Assume that Republicans voting strategically in this way knew the Powell amendment could not pass on the final vote, and keep the assumptions that precede parts (e)–(i).
(j) What rankings are possible for such a Republican who voted strategically?
(k) For which of these rankings would a strategic vote make sense?
(l) What is the least number of strategic voters that would have had to vote according to their true preferences for the original bill to pass?

7. The Democrats in the 1976 U.S. House of Representatives elected their majority leader by an elimination ballot, where the candidate receiving the least votes in any round of balloting was eliminated, and voting continued until one candidate received a majority of the votes. The results were as follows:

	Ballot 1	Ballot 2	Ballot 3
James Wright, Jr.	77	95	148
Philip Burton	106	107	147
Richard Bolling	81	93	—
John J. McFall	31	—	—

Source: Samuel Merrill, *Making Multicandidate Elections More Democratic.*

Construct or explain why it is impossible to construct orderings of the candidates, consistent with the data, so that the Condorcet winner

(a) is Wright,

(b) is Burton,

(c) is Bolling,

(d) is McFall,

(e) does not exist.

Many political observers ordered the candidates from liberal to conservative as Burton, Bolling, McFall, Wright.

(f) What is the largest number of voters whose preferences could be single-peaked with respect to this ordering of the candidates?

(g) Can you determine the Condorcet winner assuming single-peaked preferences to the greatest possible extent?

(h) Can you determine the Borda winner assuming single-peaked preferences to the greatest possible extent?

KEY TERMS

anonymity, 64

approval voting method, 48

Arrow's Impossibility Theorem, 64

Borda count, 21

Borda's method, 21

Condorcet completion process, 41

Condorcet winner, 34

head-to-head comparison, 33

independence from irrelevant alternatives, 63

neutrality, 64

nondictatorship, 63

Pareto optimality, 63

preference ranking, 10

runoff election, 8

plurality method, 4

single-peaked preference rankings, 38

strategic voting, 8

universal domain, 63

voting system, 63

weak Condorcet winner, 34

CHAPTER 1 / REVIEW TEST

1. If 454 votes are cast, what is the smallest number of votes a winning candidate can have in a six-candidate race that is to be decided by plurality?

2. Suppose there are 140 votes cast in an election among five candidates—Stein, O'Rourke, Cohen, Holt, and Massey—to be decided by plurality. After the first 100 votes are counted, the tallies are as follows:

Stein	12
O'Rourke	23
Cohen	17
Holt	29
Massey	19

   (a) What is the minimal number of remaining votes Holt needs to be assured of a win?

   (b) What is the minimal number of remaining votes Cohen needs to be assured of a win?

3. The preference rankings of a class of 24 college students for their favorite fruit are as follows:

	Number of Voters											
	2	2	1	3	2	1	3	2	3	1	2	1	1
Apples	1	1	2	3	4	3	4	2	3	4	3	4	2
Bananas	2	3	1	1	1	1	1	4	2	2	4	3	3
Grapes	3	2	3	2	2	4	3	1	1	1	1	1	4
Oranges	4	4	4	4	3	2	2	3	4	3	2	2	1

(a) Which fruit would win a plurality election?

(b) Which fruit would win a plurality election with a runoff between the top two?

(c) Which fruit has the top Borda count?

(d) Which fruit, if any, would be the Condorcet winner?

4. Suppose that a small town must decide whether to build tennis courts, a basketball court, or a baseball field. The residents of the town are polled and their preference rankings are as follows:

	Percentage of Voters					
	12	20	11	24	10	23
Tennis courts	1	1	2	3	2	3
Basketball court	2	3	1	1	3	2
Baseball field	3	2	3	2	1	1

(a)   Which choice would win a plurality election?

(b)   Which choice would win a plurality election with a runoff between the top two?

(c)   Which choice has the top Borda count?

(d)   Which choice, if any, would be the Condorcet winner?

5. A campus programming committee must decide what kind of act to book for its next engagement. The choices are a comedian, a jazz trio, a pianist, a rock band, and a classical guitarist. The committee members decide to make the decision through an approval election and the resulting ballots are as follows:

Number of Voters

	1	3	2	1	1	1	2	1	1
Comedian	✓		✓				✓	✓	
Jazz trio			✓	✓	✓		✓		✓
Pianist				✓		✓			✓
Rock band					✓		✓	✓	✓
Classical guitarist	✓					✓		✓	✓

Which act wins the vote?

6. The members of a community theater organization must vote to decide which play they would like to put on. The preference rankings of the members are as follows:

Number of Voters

	2	1	4	1	1	3	1	2
The Fantasticks	1	1	2	3	4	2	3	4
Romeo and Juliet	3	4	1	1	1	4	2	3
Our Town	4	3	3	4	2	1	1	2
Death of a Salesman	2	2	4	2	3	3	4	1

(a) Which play would win a plurality vote?

(b) Which play would win a plurality vote with a runoff between the top two finishers?

(c) Which play would win under Borda's method?

(d) Which play, if any, is the Condorcet winner?

(e) Could the three voters who ranked *Our Town* first and *The Fantasticks* second achieve a preferable outcome in an election decided by Borda's method by voting strategically if the others voted as shown in the table?

(f) Could the three voters who ranked *Our Town* first and *The Fantasticks* second achieve a preferable outcome in an election decided by the plurality method with a runoff between the top two by voting strategically if the others voted as shown in the table?

7. The eight members of a board of directors of an art museum must select a new museum head. The preference rankings of the board members and the candidates of whom they approve are included in the table.

Number of Voters

	1	1	1	2	1	2
Perella	1 ✓	1 ✓	2 ✓	3	3	3
Mintz	3	3	1 ✓	1 ✓	2	2 ✓
Zukoff	2	2 ✓	3	2	1 ✓	1 ✓

(a) Which candidate would win a plurality election?

(b) Which candidate would win a plurality election with a runoff between the top two finishers?

(c) Which candidate would win the Borda count?

(d) Which candidate, if any, is the Condorcet winner?

(e) Which candidate would win an approval vote?

(f) Could the two voters who ranked Mintz first and Zukoff second achieve a preferable outcome in an election decided by Borda's method by voting strategically if the others voted as shown in the table?

(g) Could the two voters who ranked Zukoff first and Mintz second and approved of two candidates achieve a preferable outcome in an approval election by voting strategically if the others voted as shown in the table?

8. Explain why the plurality method with a runoff between the top two finishers satisfies the Pareto optimality property.

9. Construct an example of preference rankings for an election with four candidates—A, B, C, and D—so that, in two-person races, A would defeat B, B would defeat C, C would defeat D, and D would defeat A.

SUGGESTED READINGS

Brams, Steven J. "Comparison Voting." In *Political and Related Models*, vol. 2, edited by Steven J. Brams, William F. Lucas, and Philip D. Straffin, Jr. New York: Springer-Verlag, 1983, 32–65. Negative voting, leading naturally to approval voting.

Brams, Steven J. *Game Theory and Politics*. New York: The Free Press, 1975. Strategic analysis, some mathematical, some not, mainly in the context of elections. Voting power. Numerous examples.

Brams, Steven J., and Peter C. Fishburn. *Approval Voting*. Boston: Birkhäuser, 1983. Discussion and analysis supporting approval voting. Final chapter analyzes the 1980 presidential election using linear algebra.

Fishburn, Peter C., and Steven J. Brams. "Paradoxes of Preferential Voting." *Mathematics Magazine* 56:4 (September 1983), 207–214. Examples, engagingly written.

Lind, Michael. *The Next American Nation: The New Nationalism and the Fourth American Revolution*. New York: The Free Press, 1995, 311–319. An argument in support of proportional representation, especially for the U.S. Senate.

Neufeld, John L., William J. Hausman, and Ronald B. Rapoport. "A Paradox of Voting: Cyclical Majorities and the Case of Muscle Shoals." *Political Research Quarterly* 47:2 (June 1994), 423–438. Detailed examination of the setting leading up to Exercise 15 of Section 1.3.

Niemi, Richard G., and William H. Riker. "The Choice of Voting Systems." *Scientific American* 234:6 (June 1976), 21–27. Voting methods, single-peaked preferences, and strategic voting.

Poundstone, William, *Gaming the Vote: Why Elections Aren't Fair (and What We Can Do About It)*, New York, Hill and Wang, 2008. Detailed examples of spoilers from many U.S. elections, particularly Presidential election. Descriptions of instant runoff, approval, Condorcet completion, and range voting methods.

Riker, William H. *The Art of Political Manipulation*. New Haven, Conn.: Yale University Press, 1986. Fascinating tales of how defining the issue and setting the agenda lead to favorable results in subsequent votes.

Riker, William H. *Liberalism Against Populism*. Prospect Heights, Ill.: Waveland Press, Inc., 1982. Voting methods and properties of voting systems, intermixed with lots of politics and examples.

Romer, Thomas, and Howard Rosenthal. "Voting Models and Empirical Evidence." *American Scientist* 72:5 (September–October 1984), 465–473. Geometric viewpoint of preferences, agenda control, and data on Congress.

Saari, Donald G. "Are Individual Rights Possible?" *Mathematics Magazine* 70:2 (April 1997) 83–92. Begins with a theorem of Sen pertaining to the title, which is then connected to Arrow's Impossibility Theorem.

Steen, Lynn A. "From Counting Votes to Making Votes Count: The Mathematics of Elections." Martin Gardner's column in *Scientific American* 243:4 (October 1980), 16–26B. Preferences in single-issue elections, voting methods, and voting power.

Straffin, Philip D., Jr. "The Power of Voting Blocs: An Example." *Mathematics Magazine* 50:1 (January 1977), 22–24. A nice example analyzing whether a group of delegates should organize as a bloc.

Taylor, Alan D. *Mathematics and Politics: Strategy, Voting, Power and Proof*. New York: Springer-Verlag, 1995. Includes chapters on measurement of political power and on properties of voting procedures that include plurality and Borda methods.

Cuneiform Tablets.

Before the development of coinage around 1000 B.C., people made loans and repaid them with interest. Over five thousand years ago, farmers borrowed grain for planting and repaid it with an extra measure at harvest. In ancient Babylon, financial dealings began with agricultural products such as grain and cattle as the items of exchange. Later, precious metals and then coins were exchanged. The Babylonians used mathematics to record these dealings and to compute interest payments. Many of the surviving Babylonian cuneiform clay tablets are financial records.

Mortgages for the purchase of land and pawning of personal goods in exchange for money to be repaid were established in Greece by 700 B.C. Checks payable to the bearer go back as far as Ptolemaic Egypt over two thousand years ago. Today's financial transactions are more complex, but the idea of money turned over to another in exchange for a promise of repayment with interest is unchanged.

In a complex web of financial obligations, including mortgages, car loans, credit cards, and other loans, an average household in the United States has debts totaling about $100,000 and manages to save only 5.8% of income per year. When money changes hands, it is often done through a check or an electronic transfer with no cash involved. The pulse of our society depends on this vast network of obligations governed by these transactions.

In this chapter we will learn how interest is computed. We will see how the rate of interest and the frequency with which it is compounded affects the growth of savings. We will then move from considering the growth of a single, initial deposit with interest to considering a regular series of deposits, such as in some savings plans or retirement accounts. On the other side of the coin, we conclude the chapter with a discussion of loans repaid in installments. We will learn how banks determine the size of payments on a car loan or mortgage and the total cost of paying back such loans.

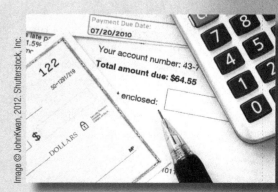

THE MATHEMATICS OF MONEY

2.1   POWERS, ROOTS, AND LOGARITHMS

Solving interest problems often requires working with exponents. Here, we review the techniques needed to solve equations involving exponents.

The product $2 \cdot 2 \cdot 2 \cdot 2$ can be written as 2^4. In a like manner, for any number x and any positive integer n, the product of n copies of x can be written as x^n. That is,

$$x^n = \underbrace{x \cdot x \cdots x}_{n \text{ times}}.$$

We also have

$$x^{-n} = \frac{1}{x^n}.$$

For instance,

$$2^{-4} = \frac{1}{2^4} = \frac{1}{16} = 0.0625.$$

The values 2^4 and 2^{-4} can be computed using the exponentiation key on your calculator. In many of the applications we will look at, the exponents will be large and a calculator will be necessary. For example, using a calculator we find that

$$(1.04)^{72} \approx 16.84226241,$$

where we have rounded the value to eight decimal places. We will use the approximation symbol $\approx$ whenever values are rounded. *When solving algebraic equations in this chapter, we will round to eight decimal places any numbers that arise in intermediate steps.* The reason for keeping all these digits is that when we round too much in an intermediate step, the error introduced may be magnified by a later step and result in a final answer that is too far off the mark. In this section, where the calculations themselves are what we are studying, we will also retain eight decimal places in our final answers. However, in the mathematics of finance problems in later sections, where we are solving for interest rates, money, or time, we will round our final answer to two decimal places. Two decimal places will normally give a final answer precise enough for business purposes, especially when solving for dollars, because we will be rounding to the nearest penny.

TECHNOLOGY *tip*

Depending on the type of calculator you are using, the exponentiation key may look like

$$\boxed{x^y} \quad \text{or} \quad \boxed{y^x} \quad \text{or} \quad \boxed{\wedge} .$$

Refer to your calculator manual to find out how to use the key on your calculator.

We will also need to compute roots of numbers. For a positive number x and a positive integer n, the nth root of x is the positive number denoted by $\sqrt[n]{x}$ whose nth power gives x. (Negative numbers have nth roots only for n odd.) We will write the nth root of x in the fractional form

$$\sqrt[n]{x} = x^{1/n}.$$

For example, because $2^3 = 8$, we see that

$$8^{1/3} = 2.$$

The exponentiation key on the calculator can be used to compute the nth root.

In some of the equations that arise in the mathematics of finance, an expression involving the unknown variable will be raised to a power that is a number. For example,

$$(1 + y)^{36} = 1.31$$

is an equation of this type. To solve for the unknown variable in this kind of equation, we will need the following rule, which holds for any positive number x and any exponents a and b:

$$(x^a)^b = x^{ab}.$$

For example,

$$(3^2)^4 = 3^8 = 6561.$$

To understand why we multiply the exponents 2 and the 4 to get the exponent 8, notice that

$$(3^2)^4 = (3^2)(3^2)(3^2)(3^2) = (3 \cdot 3)(3 \cdot 3)(3 \cdot 3)(3 \cdot 3) = 3^8.$$

We will be particularly interested in the special case of the rule that says that for x positive and any nonzero number a,

$$\boxed{(x^a)^{1/a} = x^{a \cdot (1/a)} = x^1 = x.}$$

In the next two examples, we see how this rule of exponents can be used to solve equations.

EXAMPLE 1

Solve for x in the equation $x^{15} = 1325$.

SOLUTION:

To solve for x we raise both sides to the power 1/15. Then the exponent of x will be equal to 1.

$$x^{15} = 1325$$
$$(x^{15})^{1/15} = 1325^{1/15}$$
$$x^1 = x = 1325^{1/15} \approx 1.61490777. \qquad \blacksquare$$

EXAMPLE 2

Solve for y in the equation $(1 + y)^{36} = 1.31$ where $1 + y > 0$.

SOLUTION:

We begin by raising both sides to the power 1/36, and we get

$$(1 + y)^{36} = 1.31$$
$$[(1 + y)^{36}]^{1/36} = (1.31)^{1/36}$$
$$1 + y = (1.31)^{1/36}.$$

Now we solve for y by subtracting 1 from each side of the equation and we see that

$$y = (1.31)^{1/36} - 1 \approx 0.00752895. \qquad \blacksquare$$

In interest calculations, we will often encounter equations in which we want to solve for a variable that appears in the exponent. For example,

$$(1.025)^x = 3$$

is an equation of this type. To solve for a variable in an exponent we use logarithms. We will use log, the logarithm to the base 10. For any positive number x, log x is the exponent to which 10 must be raised to give x. For instance, log 100 = 2 because $10^2 = 100$. Similarly, log (1/10) = −1 because $10^{-1} = 1/10$. Note that log x is defined only for positive x because no power of 10 is a negative number. For example, using a calculator we find that

$$\log 47 \approx 1.67209786.$$

TECHNOLOGY *tip*

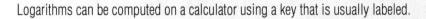

Logarithms can be computed on a calculator using a key that is usually labeled.

$$\boxed{\log x} \quad \text{or} \quad \boxed{\log}.$$

Note, however, that some books and computer software packages use log to denote the logarithm to the base e.

There are several rules of logarithms, but we need only the following rule, which holds for any positive number a and any real number b:

$$\boxed{\log(a^b) = b \log a.}$$

You may want to verify this rule with a few examples using your calculator. We will now see how this rule can be used to solve equations.

EXAMPLE 3

Solve for x in the equation $(1.025)^x = 3$.

SOLUTION:

We begin by taking the logarithms of both sides of the equation:

$$(1.025)^x = 3.$$
$$\log(1.025)^x = \log 3.$$

Next, we apply the rule previously stated and we have

$$x \log(1.025) = \log 3.$$

Finally, dividing both sides of the equation by $\log(1.025)$, we solve for x and evaluate the result on a calculator to get

$$x = \frac{\log 3}{\log(1.025)} \approx \frac{0.47712125}{0.01072387} \approx 44.49151752.$$

In this calculation, we calculated the values of log 3 and log(1.025) before performing the division. In general, it is faster to carry out the entire calculation at once on a calculator, recording only the final answer. This answer is often more accurate because the calculator may hold more decimal places in the intermediate calculations than you would write down. When the calculation in our example was done directly on a calculator, the final answer was 44.49153708.   ∎

EXAMPLE 4

Solve for t in the equation $4^{5t} + 64 = 4250$.

SOLUTION:

We first isolate the term 4^{5t} on one side of the equation:

$$4^{5t} + 64 = 4250$$
$$4^{5t} = 4250 - 64$$
$$4^{5t} = 4186.$$

Next, we take the logarithms of both sides of the equation and apply the logarithm rule to get

$$\log 4^{5t} = \log 4186$$
$$5t \log 4 = \log 4186.$$

Now we can divide both sides of the equation by both 5 and log 4 to solve for t, and we have

$$t = \frac{\log 4186}{5 \log 4} \approx 1.20313566.$$ ∎

EXAMPLE 5

Solve for x in the equation $54.73 = \dfrac{1 - (1.004)^{-12x}}{0.004}$.

SOLUTION:

Before taking logarithms, we must isolate the term $(1.004)^{-12x}$ on one side of the equation. We begin by multiplying each side of the equation by 0.004:

$$54.73 = \frac{1 - (1.004)^{-12x}}{0.004}$$
$$(54.73)(0.004) = 1 - (1.004)^{-12x}$$
$$0.21892 = 1 - (1.004)^{-12x}.$$

Next, we subtract 1 from each side to get

$$0.21892 - 1 = -(1.004)^{-12x}$$
$$-0.78108 = -(1.004)^{-12x}.$$

We now multiply both sides of the equation by −1:

$$(-1)(-0.78108) = (-1)(-(1.004)^{-12x})$$
$$0.78108 = (1.004)^{-12x}.$$

Because $(1.004)^{-12x}$ is isolated on one side of the equation, we are ready to take the logarithms of both sides of the equation and solve for x:

$$\log(0.78108) = \log(1.004)^{-12x}$$
$$\log(0.78108) = -12x\log(1.004)$$
$$x = \frac{\log(0.78108)}{-12\log(1.004)} \approx 5.15774017.$$

branching OUT

2.1

THE DEVELOPMENT OF LOGARITHMS

French mathematician and physicist Pierre Simon Laplace (1749–1827) claimed that logarithms so reduced the calculations of astronomy that they effectively doubled the life of an astronomer. Who invented (discovered is probably a better word) such an important concept? Credit goes to the Scotsman John Napier (1550–1617) and the Englishman Henry Briggs (1561–1631). Napier was a well-to-do minor lord, and included the title Master of the Mint in his resume. Briggs was brought up in modest means and continued this lifestyle as a university professor. Napier's published works on the use of logarithms and tables for their use first popularized them. Napier's "logarithm," not quite a logarithm according to today's definition, involved the constant e ≈ 2.71828183 and was somewhat unwieldy to use. Briggs' logarithm was the familiar logarithm to the base 10, for which he contributed extensive tables computed by hand, accurate to 14 decimal places. Because our number system is base 10, Briggs' logarithms were significantly easier than Napier's to use. Seventeenth-century writer William Lilly reported that the first time Napier and Briggs met, "almost one quarter of

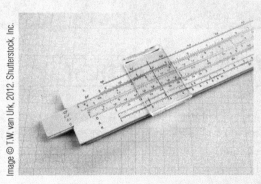

Image © T.W. van Urk, 2012. Shutterstock, Inc.

an hour was spent, each beholding the other with admiration, before one word was spoken." Logarithms led to the invention of the slide rule (in its modern form) by Englishman William Oughtred in the 1600s. The use of logarithmic tables and slide rules for approximate multiplication continued until they were supplanted by hand-held calculators in the 1970s.

EXERCISES   FOR SECTION 2.1

In each of the following exercises, solve for the unknown variable. Round your final answers to eight decimal places.

1. $x^9 = 568$

2. $x^{17} = 39$

3. $y^{50} = 7.23$ where $y > 0$

4. $z^5 = 358$

5. $x^{12} = 1570$ where $x > 0$

6. $t^{72} = 1.35$ where $t > 0$

7. $3^x = 782$

8. $5^x = 6026$

9. $(1.73)^{-z} = 8$

10. $6^{-x} = 0.32$

11. $9^t = 0.5$

12. $(1.047)^x = 9.2$

13. $8^{2x} = 49$

14. $6^{5y} = 0.39$

15. $(7.53)^{-4x} = 0.249$

16. $9^{-8x} = 104$

17. $(1 + x)^{108} = 7.3$ where $1 + x > 0$

18. $(1 + x)^{12} = 68$ where $1 + x > 0$

19. $(3 + t)^4 = 82$ where $3 + t > 0$

20. $(x - 2)^{72} = 4.32$ where $x - 2 > 0$

21. $4 + 2^t = 6520$

22. $(6.4)^x - 4 = 20$

23. $4^{-12z} + 1 = 1.75$

24. $1 + 7^{9z} = 6$

25. $(1.5)^{2x} - 3 = 8.6$

26. $2 + (9.2)^{-8x} = 2.32$

27. $186 = \dfrac{(1.02)^{4x} - 1}{0.02}$

28. $57.5 = \dfrac{(1.003)^{12t} - 1}{0.003}$

29. $72.3 = \dfrac{1 - (1.01)^{-12y}}{0.01}$

30. $2.41 = \dfrac{1 - (1.015)^{-4x}}{0.015}$

2.2   SIMPLE INTEREST

When we invest money in a bank account, the bank pays us interest. The most basic form of interest is simple interest. With **simple interest**, the interest is paid only at the end of a specified period of time and is paid only on the amount initially deposited, called the **principal**. The interest paid depends on the principal, the interest rate, and the length of time the account is held.

Suppose $200 is deposited into a simple interest account at 5% interest per year for 1 year. Then the amount of interest I is 5% of $200, which is given by

$$I = (200)(0.05) = \$10.$$

If the account is held for 6 years, then the interest paid is 6 times the amount of interest paid for 1 year and is given by

$$I = (200)(0.05)(6) = \$60.$$

Following this line of reasoning, we see that in general

simple interest = principal · rate · time.

We can now write a general algebraic formula for simple interest.

Simple Interest Formula

$$I = Prt \quad \text{where} \begin{cases} I = \text{interest} \\ P = \text{principal} \\ r = \text{annual interest rate} \\ t = \text{time (in years)} \end{cases}$$

The interest rate r we use in the formula is the annual interest rate. It is important to note that the interest rate r must be written as a decimal. If you are given the interest rate as a percentage, you convert it to a decimal by dividing by 100. For example, if the annual rate is 9.6%, then $r = 9.6/100 = 0.096$.

EXAMPLE 1 *Simple Interest*

If $3500 is invested at the simple interest rate of 7% per year, how much interest will be earned if the investment is held for 4 years?

SOLUTION:

Here the principal is $P = 3500$, the annual interest rate is $r = 0.07$, and $t = 4$ years. Using the simple interest formula $I = Prt$, we find that the interest earned is

$$I = (3500)(0.07)(4) = \$980.$$

EXAMPLE 2 *Simple Interest*

Suppose $2700 is deposited into an account earning simple interest of 4.8% annually. If the account is closed at the end of 15 months, how much interest will the bank pay?

SOLUTION:

Because the time is given in months rather than years, we first convert the months to years. We do this by dividing by 12, and we find that $t = 15/12 = 1.25$ years. Now we can use the simple interest formula $I = Prt$ with $P = 2700$, $r = 0.048$, and $t = 1.25$, and we see that the interest paid is

$$I = (2700)(0.048)(1.25) = \$162. \qquad \blacksquare$$

For any kind of investment, we are interested in what it will grow to be worth in the future. Appropriately, this amount is called the **future value** of the investment. The future value of any investment will depend on how much is invested and on the rate and method by which interest is paid. It will also depend on whether the investment is made in one lump sum or by several deposits over time. For a lump-sum investment earning simple interest we can easily compute its future value by adding the interest to be earned to the original principal invested. Using the letter F to denote the future value, we see that the future value of a simple interest investment is given by

F = principal + interest = principal + principal · rate · time.

Letting P denote the principal, I interest, r the rate, and t time, we see that

$$F = P + I = P + Prt = P(1 + rt).$$

Conversely, if we have a particular future value F in mind, we might ask what amount, P, would have to be deposited today in order to achieve a future value of F after t years of earning interest at a rate r. We call this amount P the **present value** corresponding to the future value F. For a simple interest investment, the present value is simply the principal P that will result in a future value F. Therefore, in the simple interest future value formula $F = P(1 + rt)$, we can think of P as either the principal or the present value depending on the particular viewpoint of the problem we are trying to solve.

Simple Interest Future Value Formula

$F = P(1 + rt)$ where $\begin{cases} F = \text{future value after } t \text{ years} \\ P = \text{principal or present value} \\ r = \text{annual interest rate} \\ t = \text{time (in years)} \end{cases}$

EXAMPLE 3 *Finding the Future Value*

If you put $4250 in an account today that earns simple interest at a rate of 6.53% per year, what will the investment be worth 8 years from now?

SOLUTION:

We want to find the future value F of the investment after 8 years. Using the simple interest future value formula with the principal $P = 4250$, $r = 0.0653$, and $t = 8$, we see that after 8 years the investment will be worth

$$F = 4250(1 + (0.0653)8) = 4250(1 + 0.5224) = 4250(1.5224) = \$6470.20. \quad \blacksquare$$

EXAMPLE 4 *Finding the Present Value*

Find the amount of money you must deposit in a simple interest account paying 4.3% annually so that the account will be worth $5000 in 3 years.

SOLUTION:

In this example we want to find the present value, P, in the simple interest future value formula given that $F = 5000$, $r = 0.043$, and $t = 3$. Writing the formula and substituting the known values, we have

$$F = P(1 + rt)$$
$$5000 = P(1 + (0.043)3)$$
$$5000 = P(1 + 0.129)$$
$$5000 = P(1.129).$$

Solving for P gives

$$P = \frac{5000}{1.129} \approx \$4428.70.$$

Note that in the last step the value of P was rounded to two decimal places. $\blacksquare$

When a loan specifies simple interest, you can compute the interest due or the balance due on the loan using the formulas for simple interest or future value in the usual way. Such calculations are shown in the next two examples.

EXAMPLE 5 *Simple Interest Loan*

A man borrows $300 from his parents to be repaid in 100 days at a simple interest rate of 5% annually. How much will he need to repay them 100 days later?

He will owe his parents the future value of $300 after 100 days, so we use the simple interest future value formula $F = P(1 + rt)$. Because t must be given in years, we have to convert 100 days to years by dividing by 365. We get $t = 100/365 \approx 0.27397260$ after rounding to eight decimal places. We have $P = 300$ and $r = 0.05$, so the future value is given by

$$F \approx 300(1 + (0.05)(0.27397260)) = 300(1 + 0.01369863)$$
$$= 300(1.01369863) \approx \$304.11,$$

and we see that he will owe $304.11 at the end of 100 days. ■

EXAMPLE 6 *Payday Loan*

Payday loans are short-term cash advances due on your payday. These loans are easily obtained through many internet companies as well as through local specialized lenders. In February 2012, EZPayDayCash.com was charging a fee of $60 for a 7-day cash advance of $200. What is annual rate of simple interest on this loan?

In this case, the present value is the amount borrowed $P = 200$, and the future value is the amount owed after 7 days, which is the $200 borrowed plus the $60 fee, giving $F = \$200 + \$60 = \$260$. The time is $t = 7$ days. Because t must be given in years, we have to convert 7 days to years by dividing by 365. We get $t = 7/365 \approx 0.01917808$ after rounding to eight decimal places. To find the rate r, we substitute into the simple interest future value formula and get

$$260 \approx 200(1 + r(0.01917808)).$$

To solve for r, we first isolate $r(0.01917808)$ by dividing both sides by 200 and then subtracting 1 from each side:

$$\frac{260}{200} \approx 1 + r(0.01917808)$$
$$1.3 \approx 1 + r(0.01917808)$$
$$1.3 - 1 \approx r(0.01917808)$$
$$0.3 \approx r(0.01917808).$$

Now, solving for r gives

$$r \approx \frac{0.3}{0.01917808} \approx 15.64285893.$$

We multiply by 100 to covert the rate r to a percent and find

$$r = 15.64285893 \cdot 100\% \approx 1564.29\%.$$

So we see that a staggering interest rate of 1564.29% is being charged for this payday loan. ■

branching OUT

2.2

THE ORIGIN OF THE WORD AND SYMBOL FOR PERCENT

The word percent, meaning per hundred, goes back to the Latin term per centum. Early twentieth-century scholar D. E. Smith traced the % symbol back to 1425 in Italy, where per cento meant per hundred. The cento, or hundred, was written $\tilde{c}$. The unknown author who introduced the precursor of the modern % notation wrote $\mathbf{P}\,c\!\!\!\sim^{\!\!o}$. Around 1650, the $c\!\!\!\sim^{\!\!o}$ became $\frac{0}{0}$. When the fraction $\frac{a}{b}$ was typeset as a/b some two hundred years ago, per cent became 0/0 or stylized as %, and the $\mathbf{P}$ disappeared along the way. There is a similar, though obscure, notation ‰, meaning per mille, or per thousand.

Image © Krol, 2012. Shutterstock, Inc.

The interest rate of 1564.29% seen in the last example is typical of the shocking interest rates charged by payday loan companies. Why do these businesses get away with this? They are not taking on very much risk because the borrowers must allow the lender to automatically deduct the repayment from their paycheck or, in other instances such as pawn shops, must leave some kind of item or items of value as collateral that can be sold if the loan is not repaid. Businesses like these can charge high interest rates because this type of transaction is easy and convenient for their customers who, most likely, do not have ready access to loans at lower rates.

EXERCISES FOR SECTION 2.2

The interest rates given in the following exercises are annual rates. Round all dollar amounts in the answers to the nearest cent.

In Exercises 1–6, find the simple interest I earned on the given principal for the time period and interest rate specified.

1. $P = \$4000$, $r = 5\%$, $t = 3$ years
2. $P = \$2300$, $r = 4\%$, $t = 5$ years

3. $P = \$752$, $r = 12.1\%$, $t = 8$ years
4. $P = \$20{,}000$, $r = 6.25\%$, $t = 2$ years
5. $P = \$35{,}200$, $r = 2\%$, $t = 7$ months
6. $P = \$85$, $r = 3.5\%$, $t = 14$ months

In Exercises 7–10, find the future value F of the given principal earning simple interest for the time period and interest rate specified.

7. $P = \$320$, $r = 3\%$, $t = 4$ years
8. $P = \$7123$, $r = 10.5\%$, $t = 5$ months
9. $P = \$51{,}225$, $r = 4.35\%$, $t = 50$ days
10. $P = \$8000$, $r = 7\%$, $t = 2$ years

In Exercises 11–14, find the present value P under simple interest given the future value, time period, and interest rate specified.

11. $F = \$2000$, $r = 9\%$, $t = 6$ months
12. $F = \$830$, $r = 8.4\%$, $t = 3$ years
13. $F = \$90{,}000$, $r = 14.2\%$, $t = 4$ years
14. $F = \$4400$, $r = 6\%$, $t = 10$ weeks

In Exercises 15 and 16, find the time t, in years, for the given principal to reach the future value under simple interest for the interest rate specified.

15. $P = \$300$, $F = \$400$, $r = 6.2\%$ 16. $P = \$20{,}150$, $F = \$29{,}000$, $r = 8\%$

In Exercises 17 and 18, find the annual interest rate r, for the given principal to reach the future value under simple interest for the time period specified.

17. $P = \$123$, $F = \$140$, $t = 10$ months   18. $P = \$1500$, $F = \$1850$, $t = 2$ years

19. Six years ago, a woman made an investment paying 7.5% simple interest. If her investment is now worth $3245, how much did she originally invest?

20. At Vanderbilt University in 2012, if a student did not pay tuition and fees on time, the student's account would be assessed a monthly late-payment fee of 18% annual simple interest. If a student owed $13,943 and the payment was two months late, what is the total amount the student would have to pay?

21. Many investors wish they had invested in Apple stock when it first went public on December 12, 1980. If you had invested $8000 in Apple stock on that date, then the stock would have risen in value to $1,139,898.18 by December 12, 2011. What was the rate of return, figured as an annual simple interest rate, for this investment?

22. The Continental Congress borrowed $3.8 million in 1777 at 6% interest. How much simple interest would it have paid by 1782, when it was unable to continue making payments?

23. A man received a salary bonus of $1000. He needs $1150 for the down payment on a car. If he can invest his bonus at 8.25% simple interest, how long must he wait until he has enough money for the down payment?

24. A woman borrowed $75 from her parents, agreeing to pay them back in 9 months with simple interest. If the woman paid her parents $100 at the end of 9 months, what annual simple interest rate did she pay?

25. In 44 B.C., Brutus, the man who murdered Julius Caesar, made a loan to the city of Salamis. He charged an interest rate of 48%, even though 12% was the maximum legal rate. How much simple interest would Brutus make on a loan of 3000 talents for 3 years?

26. Suppose some municipal bonds pay 6.2% simple interest. How much should you invest in the bonds if you want them to be worth $5000 in 10 years?

27. The state of Florida limits pawn shop interest rates to a maximum of 300% simple interest. Suppose you borrow $50 from a pawn shop charging 300% simple interest. How much would you owe if you paid back the loan after 7 months?

28. If $32,000 is invested at 4.83% simple interest, how long will it take for the investment to be worth $40,000?

29. Suppose a man borrows $500 from a friend agreeing to pay it back in 2 years with 6% simple interest. How much will the man owe his friend at the end of the 2 years?

30. When people arrested for a crime do not have the money to post bail, they often call on a bail bondsman to put up the bail money. Bail bondsmen get their money back once the court proceedings of the arrested person have concluded. Laws in the state of Tennessee are fairly typical, allowing the bondsman to charge a premium of 10% of the bond and an additional $25 fee. If a person's bond is $10,000 and the case is over in 3 months, what is the annual simple interest rate on what is essentially a $10,000 loan for 3 months? Consider the combined amount of the 10% premium and the $25 fee to be the interest paid.

31. In February 2012, PayDayMax.com was charging a fee of $93.10 for a 14-day payday loan of $500. What is annual rate of simple interest on this loan?

2.3 COMPOUND INTEREST

Many kinds of investments, including a typical savings account in a bank, earn interest that is compounded at regular intervals. The interest is paid on the initial deposit as well as on any previous interest payments credited to the account. This kind of interest is called **compound interest.** If banks only offered simple interest, then depositors could increase their interest earnings by withdrawing the money, drawing interest, and redepositing both the original principal and the interest earned, so that interest is paid on any previous interest earned. This would, at the very least, lead to a lot of extra work for the banks, so they offer compound interest.

As you might guess, the formula for finding the future value of an account with compounded interest is less obvious than for a simple interest account, so we begin by considering a simple example. Suppose you deposit $300 into an account earning 5% interest compounded annually and leave the account open for 4 years without any further deposits. At the end of the first year, interest will be paid on the $300, so the total amount in the account will be

$$300 + 300(0.05) = 300(1 + 0.05) = 300(1.05).$$

Notice that adding 5% interest is the same as multiplying by (1.05). At the end of the second year, 5% interest is paid again. Because the balance at the start of the second year is 300(1.05) = $315 and because paying 5% interest is equivalent to multiplying by (1.05), we see that, after interest is paid, the amount in the account at the end of 2 years will be

$$[300(1.05)](1.05) = 300(1.05)^2.$$

At the end of 3 years, interest will be paid on the balance of $300(1.05)^2 = \$330.75$ from the previous year, so the amount in the account will be

$$[300(1.05)^2](1.05) = 300(1.05)^3.$$

Finally, following this line of reasoning, we find that at the end of 4 years the amount in the account will be

$$[300(1.05)^3](1.05) = 300(1.05)^4 \approx \$364.65,$$

where the final amount was rounded to the nearest cent. This example suggests the formula for the future value of an account earning interest compounded annually. If P is the principal invested and r is the annual rate of interest compounded annually, then the future value F after t years, where t is an integer, is given by $F = P(1 + r)^t$.

In the general situation, interest may be credited more often than once a year. The interval of time between interest payments is called a **period.** Let n be the number of periods per year. For example, if interest is compounded monthly, then $n = 12$. The interest that is paid each period is called the periodic interest.

If the annual rate of interest is r and interest is compounded n times per year, then the **periodic interest rate** is given by

$$\text{periodic interest rate} = \frac{r}{n}.$$

For example, if the annual interest rate is 4.8% and the interest is compounded monthly, then the periodic interest rate is given by

$$\frac{r}{n} = \frac{0.048}{12} = 0.004.$$

Suppose P dollars is invested in a compound interest account. As in the case of simple interest, we call this initial amount P the present value or principal. Following the ideas from our example, we see that after one period the amount in the account is given by

$$P\left(1 + \frac{r}{n}\right),$$

and, in general, after k periods the amount in the account is given by

$$P\left(1 + \frac{r}{n}\right)^{k}.$$

Because there are n periods each year, we see that the total number of periods is given by $k = nt$, where t is the number of years the account is held. Combining these ideas, for nt an integer, we have the following general compound interest formula.

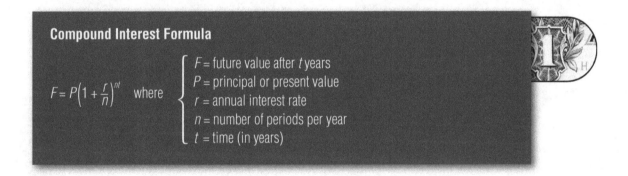

Compound Interest Formula

$$F = P\left(1 + \frac{r}{n}\right)^{nt} \quad \text{where} \quad \begin{cases} F = \text{future value after } t \text{ years} \\ P = \text{principal or present value} \\ r = \text{annual interest rate} \\ n = \text{number of periods per year} \\ t = \text{time (in years)} \end{cases}$$

To remember this formula, think of it as

future value = principal · (1 + periodic rate)^{total number of periods}.

To see the effect of compounding interest, we look at a graph. In Figure 2.1, the amount in an account with an initial deposit of $1000 that earns 8% interest is graphed both for the case of simple interest and for the case where the interest is compounded monthly. The graph for simple interest is a straight line, whereas the graph for the accrued value of the account earning compounded interest is curved upward. This is always true. The increasing slope of the compound interest curve shows us why compound interest always eventually outpaces simple interest, even if the simple interest rate is much higher.

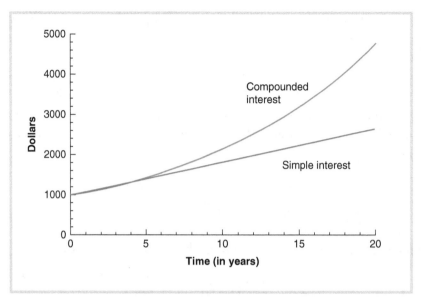

FIGURE 2.1 Amount in an account with an initial deposit of $1000 at 8% interest with interest compounded monthly and with simple interest.

EXAMPLE 1 *Finding the Future Value*

Suppose $2400 is deposited in an account earning 5.54% interest compounded quarterly. How much will be in the account after 9 years?

SOLUTION:

Because interest is compounded quarterly, $n = 4$. We also have $P = 2400$, $r = 0.0554$, and $t = 9$. Using the compound interest formula, we find that the amount in the account after 9 years is given by

$$F = 2400 \left(1 + \frac{0.0554}{4}\right)^{4 \cdot 9} = 2400(1.01385)^{36} \approx \$3937.90,$$

where the final answer has been rounded to the nearest penny. ■

The more frequently interest is compounded, the faster the account grows. This principle is illustrated in Example 2.

EXAMPLE 2 *Comparing Compounding Periods*

Find the future value of a $2000 investment after 6 years if it is earning an annual interest rate of 7% compounded

(a) annually,
(b) daily.

SOLUTION:

(a) For interest compounded annually, $n = 1$, so after 6 years

$$F = 2000(1 + 0.07)^6 = 2000(1.07)^6 \approx \$3001.46.$$

(b) When the interest is compounded daily, $n = 365$, and we have

$$F = 2000\left(1 + \frac{0.07}{365}\right)^{365 \cdot 6} \approx 2000(1.00019178)^{2190} \approx \$3043.80.$$

With daily compounding, the future value is $42.34 more than the future value of the investment when it is compounded annually. ■

If we had rounded the amount $1 + \frac{0.07}{365}$ to 1.000192 in the last calculation, we would have obtained $F \approx \$3045.26$, whereas rounding to 1.0002 would give $F \approx \$3099.07$. You get the most accurate answer by using all the decimal places you have available in each step of the calculations. Recall our conventions in this chapter: When solving algebraic equations, we will round any numbers that arise in intermediate steps to eight decimal places, and we will round our final answers in the mathematics of finance problems—where we are solving for interest rates, money, or time—to two decimal places.

In the last example, we saw that when the interest was compounded more frequently, the interest earned over time was greater. Because the actual interest earned on an account depends on the annual rate and on how often the interest is compounded, it would be nice to have a single number that allows us to compare different accounts. There is such a number and it is commonly called the **annual percentage yield** or **APY.** The APY is the actual percentage increase in the account over a 1-year period. Banks often advertise their interest rates in terms of APY. From the compound interest formula, we see that in a 1-year period the principal P is increased by a factor of

$$\left(1 + \frac{r}{n}\right)^{n \cdot 1} = \left(1 + \frac{r}{n}\right)^{n}.$$

For instance, if the annual rate is 6% and the interest is compounded monthly, the principal P is increased by a factor of

$$\left(1 + \frac{0.06}{12}\right)^{12} = (1.005)^{12} = 1.0617.$$

Therefore, we see that the annual percentage yield is 0.0617 or 6.17%. Following this idea, we have the following general formula:

Annual Percentage Yield (APY)

The annual percentage yield of an account with an annual interest rate r compounded n times per year is given in decimal form by

$$\text{APY} = \left(1 + \frac{r}{n}\right)^{n} - 1.$$

The APY is sometimes referred to as the annual effective yield or annual equivalent yield.

EXAMPLE 3 *APY*

For an account with an annual interest rate of 8%, find the annual percentage yield when it is compounded

(a) quarterly,
(b) monthly,
(c) daily.

SOLUTION:

(a) To find the APY when the interest is compounded quarterly, we use our formula for the annual percentage yield with $r = 0.08$ and $n = 4$. We see that

$$\text{APY} = \left(1 + \frac{0.08}{4}\right)^4 - 1 = (1.02)^4 - 1 = 0.08243216 \approx 8.24\%.$$

(b) Similarly, when the interest is compounded monthly, we find that

$$\text{APY} = \left(1 + \frac{0.08}{12}\right)^{12} - 1 \approx (1.00666667)^{12} - 1 \approx 0.08299955 \approx 8.30\%.$$

(c) For interest compounded daily,

$$\text{APY} = \left(1 + \frac{0.08}{365}\right)^{365} - 1 \approx (1.00021918)^{365} - 1 \approx 0.08327833 \approx 8.33\%. \qquad \blacksquare$$

In Example 3, we saw that going from compounding monthly to daily resulted in a smaller increase in the APY than when we went from compounding quarterly to monthly. It turns out that going from compounding daily to compounding every hour, minute, second, or even "continuously" does not have much effect on the APY.

If you want a single deposit, earning compound interest at a fixed rate, to be worth some particular future value after a certain number of years, how much should you deposit? This is a question of finding the present value P in the compound interest formula. We see this idea in the next example.

EXAMPLE 4 *Finding the Present Value*

A couple wants to invest some money for their child's college education. If they want to have $80,000 in 10 years and the investment pays 6.5% compounded monthly, how much should they deposit?

SOLUTION:

In this situation, we find the present value P corresponding to future value $F = 80,000$ when $r = 0.065$, $t = 10$, and $n = 12$. Substituting into the compound interest formula gives

$$80,000 \approx P\left(1 + \frac{0.065}{12}\right)^{12 \cdot 10}$$
$$80,000 \approx P(1.00541667)^{120}$$
$$80,000 \approx P(1.91218451).$$

and solving for P gives

$$P \approx \frac{80,000}{1.91218451} \approx \$41,836.97.$$

This is a rather large amount of money to deposit at one time. A more typical way to save for college tuition is to make several smaller deposits over a period of time. We will look at systematic savings plans of this type in the next section. ■

A CLOSER *look*

THE DIMINISHING EFFECT OF COMPOUNDING MORE FREQUENTLY

What effect does compounding more often than daily have on the APY? Let's see by comparing the APYs corresponding to an annual rate of 10% for interest compounded quarterly, monthly, daily, and every hour, minute, or second. (To avoid rounding problems, all intermediate calculations here were carried to 20 decimal places.)

Length of Period	APY (percent)
Quarter	10.38128906
Month	10.47130674
Day	10.51557816
Hour	10.51702873
Minute	10.51709076
Second	10.51709179

We see that the APY does not increase much when interest is compounded more often than daily. By letting the number of periods n get larger and larger, we have the idea of compounding continuously. The formula for the APY that results in this case is $e^r - 1$, where r is the annual interest rate and $e \approx 2.71828183$ is a special number that arises naturally in many areas of mathematics. In the case of an annual rate of 10%, the APY when interest is compounded continuously is given by

$$e^{0.10} - 1 \approx 0.1051709181 = 10.51709181\%,$$

which is only marginally higher than the APY when compounding every second.

In some situations, the amount you have to invest is fixed but you may have in mind a goal of how much you want the investment to be worth in the future. The question is to find the interest rate required to reach that goal. In choosing investments, we often have some choice of interest rates, but with higher rates there may be some form of risk assumed by the investor. In the next example, we see how to solve for the interest rate required to meet a financial goal.

EXAMPLE 5 *Finding the Interest Rate*

If \$2500 is invested with interest compounded quarterly, what annual interest rate is needed for the investment to be worth \$3000 in 15 months?

SOLUTION:

We must write the time t in years, so we divide 15 months by 12 to get

$$t = \frac{15}{12} = 1.25 \text{ years.}$$

We also have $P = 2500$, $F = 3000$, and $n = 4$. Using the compound interest formula, we get

$$3000 = 2500\left(1 + \frac{r}{4}\right)^{4 \cdot 1.25}$$

$$3000 = 2500\left(1 + \frac{r}{4}\right)^{5}.$$

Dividing each side by 2500, we have

$$\frac{3000}{2500} = \left(1 + \frac{r}{4}\right)^{5}$$

$$1.2 = \left(1 + \frac{r}{4}\right)^{5}.$$

Now we raise both sides to the power 1/5,

$$(1.2)^{1/5} = \left[\left(1 + \frac{r}{4}\right)^{5}\right]^{1/5}$$

$$1.03713729 \approx \left(1 + \frac{r}{4}\right),$$

and solving for r gives

$$0.03713729 \approx \frac{r}{4}$$

$$r \approx 4(0.03713729)$$

$$r \approx 0.1485 = 14.85\%.$$

We see that an annual interest rate of about 14.85% will make the investment grow to the desired amount. ∎

Next we will see how we can use logarithms to solve for the time t in a problem involving compound interest.

EXAMPLE 6 *Finding Time*

If $1600 is deposited into an account earning 6% compounded monthly, how long will it take for the account to be worth $2500?

SOLUTION:

In this case, we know that $P = 1600$, $F = 2500$, $r = 0.06$, and $n = 12$. Substituting into the compound interest formula gives

$$2500 = 1600\left(1 + \frac{0.06}{12}\right)^{12t}$$

$$2500 = 1600(1.005)^{12t}.$$

Because t appears in an exponent, we use logarithms to solve for it. We first divide both sides by 1600 to isolate the term $(1.005)^{12t}$:

$$\frac{2500}{1600} = (1.005)^{12t}$$

$$1.5625 = (1.005)^{12t}.$$

Next we take the logarithm of each side of the equation, obtaining

$$\log(1.5625) = \log(1.005)^{12t}$$

$$\log(1.5625) = 12t \log(1.005).$$

Finally, solving for t gives

$$t = \frac{\log(1.5625)}{12\log(1.005)} \approx 7.46 \text{ years.}$$

∎

branching OUT

THE HIGH COST OF INTEREST: THE INTER CHANGE BANK JUDGMENT

Shock waves went through the small Swiss canton of Ticino when, on October 3, 1994, it was ordered by a New York state court to pay more than $125 billion to an American family who claimed to have lost money when the local Inter Change Bank failed in the 1960s. The roots of the case date back to 1961, with the death of Seattle real estate investor Sterling Granville Higgins. In 1989, lawyers for Mr. Higgins' brother Robert Higgins filed suit in New York, claiming that in 1966 the estate of Sterling Higgins deposited $600 million in options on oil and mineral deposits in Venezuela with Inter Change Bank. The plaintiffs claimed that when the bank went bankrupt a year later many of the estate documents were lost. They claimed the bank had agreed to pay an extraordinary interest rate of 1% per week on the investment.

Because the Swiss government bankruptcy agencies that liquidated the bank never showed up in court, state Justice Gerald Held ruled in the estate's favor and ordered the bank to pay $125,444,300,221, corresponding to the value in 1994 of a $600 million investment made in 1966 with an annual percentage yield of about 21%. If the interest had actually been calculated at 1% per week compounded weekly, the judgment would have been for over $1000 trillion.

Ticino, with a population of only about 300,000 and annual government budget of about $2.1 billion, was in no position to pay the judgment and appealed the decision. The circumstances surrounding the case raised many questions. Both of the Higgins brothers had spent time in jail. Sterling died in San Quentin Prison, where he was serving time for writing bad checks. Robert had been convicted of fraud. At the time of Sterling Higgins' death, his net worth according to papers filed with the court was less than $5000. Swiss officials claimed that it was impossible that $600 million was deposited in 1966 because the bank books did not show anything near that in total value. They also claimed they were never told to appear in court and believed they were not subject to a judgment from a U.S. state court. The case was moved to U.S. District Court in Brooklyn, and on February 28, 1996, much to the relief of Ticino and the Swiss government, the judgment was overturned.

Image © kor, 2012. Shutterstock, Inc.

In the last example, we found that the account would be worth $2500 in about 7.46 years, which translates to 7 years, 5.52 months. However, if the interest is only paid at the end of the month and only on money left until the end of the month, then we would have to wait 7 years, 6 months to have at least $2500.

Because interest is only paid at the end of each period, the formula is only exactly correct at the end of each period in this case. We are not concerned with this when solving for time, because our answers give a reasonable approximation to the answer we are seeking. We will take this approach whenever we are solving for time throughout this chapter. Note, however, that with some investments, if the account is closed in the middle of a period, interest will be paid even for the fraction of the period the money was invested. In such cases, the compound interest formula always gives a future value very close to the closing balance.

Inflation

Inflation is the increase in the prices of goods and services. If a 10% inflation rate holds for 3 years running, the cost of a $10 lunch increases to $11 after 1 year, then to $12.10 after 2 years, and to $13.31 after the third year. Inflation works just like compounding of interest, and we can use the idea of compound interest to calculate the effect of inflation on prices and salaries. In this case, we consider the annual inflation rate as the rate of interest and view the increases in salaries and prices as being compounded annually. This idea is illustrated in our next two examples.

EXAMPLE 7 *Inflation*

Suppose the rate of inflation for the next 5 years is projected to be 4% per year. How much would you expect a car that costs $23,000 today to cost in 5 years?

SOLUTION:

An inflation rate of 4% per year has the same effect as paying interest of 4% per year compounded annually, where the price of the car today is the present value P and the price it will be in 5 years is the future value F. Therefore, we can use the compound interest formula with $P = 23,000$, $r = 0.04$, $t = 5$, and $n = 1$. Evaluating F we get

$$F = 23,000\left(1 + \frac{0.04}{1}\right)^{1 \cdot 5}$$
$$F = 23,000(1.04)^5$$
$$F \approx 23,000(1.21665290)$$
$$F \approx \$27,983.02.$$

Therefore, we should expect the car to cost $27,983.02 or roughly $28,000 in 5 years. ∎

branching OUT

HYPERINFLATION

Lenin stated in 1910 that "The surest way to overthrow an established social order is to debauch its currency." In a 1919 book, the famous British economist John Maynard Keynes paraphrased Lenin and went on to say "The process [inflation] engages all the hidden forces of economic law on the side of destruction, and does it in a manner which not one man in a million is able to diagnose." Whereas the highest rate of inflation in the United States since 1920 was 14.4% in 1947, other countries have suffered from runaway inflation, or *hyperinflation,* where prices might double or triple in a month, or even worse. If prices triple every month, they would increase by a factor of $3^{12} = 531,441$ (or 53,144,100%) over the course of a year. Such inflation rates rarely last beyond a few months, but even much lower increases take a great toll over the course of several years. For instance, hyperinflation almost inevitably results in shortages of consumer goods. It is most common in developing countries with heavy national debts. Simplistically speaking, the government simply prints the money it needs to remain solvent. Unfortunately, such action diminishes the purchasing power of all the money in circulation. Weimar Germany, in the aftermath of World War I, provides an early example. Before the war, the combined value of all mortgages in the country was on the order of forty billion marks. When inflation peaked in 1923, forty billion marks might buy a loaf of bread. To combat this problem, mark notes worth billions were put into circulation.

A League of Nations report on the inflation throughout Europe in the 1920s noted that "Inflation is the one form of taxation which even the weakest government can enforce, when it can enforce nothing else." An inflation rate averaging over 10% *per day* in Hungary in 1945 resulted in what is possibly the largest denomination of currency ever circulated, the 100 quintillion (100,000,000,000,000,000,000) pengo note issued in 1946.

A recent case of astonishing hyperinflation occurred in the African country of Zimbabwe. When Zimbabwe first became an independent country in 1980, a Zimbabwe dollar was worth about $1.25 in US dollars. During the 1980s and until the late 1990s, inflation was held to double digits. However, beginning in the late 1990s food production in the country was severely reduced due to the country's policy of land confiscation from commercial farmers. In order to fund the government in the face of the resulting decreased tax revenue, the Reserve Bank of Zimbabwe began printing

10 billion mark note from Germany.

Image © Fedor Kondratenko, 2012. Shutterstock, Inc.

more money and the inflation rate rose to triple digits by 2001. By the end of 2008, prices were doubling every day. In January 2009, Zimbabwe issued a 100 trillion dollar note, which at the time was worth only about thirty US dollars. In the midst of the raging hyperinflation, teachers and civil workers went on strike due to the fact that their salaries, which were in trillions of dollars, could not cover even the cost of taking a bus to work. Zimbabwe finally saved itself from this disastrous situation by suspending the use of it own currency and allowing the use of foreign currency. It is easy to see why the civil, economic, and political instability caused by severe inflation steers all but the most nearsighted and desperate of politicians away from monetary recklessness.

100 trillion dollar note from Zimbabwe.

EXAMPLE 8 *Inflation*

Based on the Consumer Price Index, the rate of inflation in the United States between 1990 and 2011 was about 2.5% per year. What salary in 1990 would be equivalent to a $40,000 salary in 2011?

SOLUTION:

To solve this, we view the problem from a 1990 perspective. Then the salary of $40,000 in 2011 is the future value F, and the equivalent salary in 1990 is the corresponding present value P under an interest rate of 2.5% compounded annually. Because there are 21 years between 1990 and 2011, we have $t = 21$. Substituting $F = 40,000$, $r = 0.025$, $t = 21$, and $n = 1$ into the compound interest formula and solving for P, we have

$$40,000 = P\left(1 + \frac{0.025}{1}\right)^{1 \cdot 21}$$

$$40,000 = P(1.025)^{21}$$

$$40,000 \approx P(1.67958185)$$

$$P \approx \frac{40,000}{1.67958185} \approx \$23,815.45.$$

EXERCISES FOR SECTION 2.3

In Exercises 1–8, find the future value F of the given principal under compound interest for the time period and interest rate specified.

1. $P = \$6000$, $t = 7$ years, $r = 4\%$ compounded annually
2. $P = \$35,000$, $t = 10$ years, $r = 7.3\%$ compounded annually
3. $P = \$7250$, $t = 3$ years, $r = 10\%$ compounded quarterly
4. $P = \$900$, $t = 20$ years, $r = 2.5\%$ compounded monthly
5. $P = \$2356$, $t = 9$ months, $r = 5.7\%$ compounded monthly
6. $P = \$50,000$, $t = 6$ years, $r = 3.5\%$ compounded quarterly
7. $P = \$544$, $t = 15$ years, $r = 5\%$ compounded daily
8. $P = \$7000$, $t = 2$ years, $r = 11.25\%$ compounded daily

In Exercises 9–12, find the present value P given the future value, time period, and interest rate specified.

9. $F = \$5400$, $t = 4$ years, $r = 6.5\%$ compounded annually
10. $F = \$16,200$, $t = 18$ months, $r = 3\%$ compounded quarterly
11. $F = \$185$, $t = 2$ years, $r = 10.3\%$ compounded daily
12. $F = \$14,000$, $t = 24$ years, $r = 6\%$ compounded monthly

In Exercises 13–16, find the time t, in years, for the given principal to reach the given future value at the interest rate specified.

13. $P = \$1200$, $F = \$2000$, $r = 5.8\%$ compounded monthly
14. $P = \$348$, $F = \$500$, $r = 1.53\%$ compounded daily
15. $P = \$10,000$, $F = \$40,000$, $r = 9\%$ compounded quarterly
16. $P = \$960$, $F = \$1400$, $r = 13\%$ compounded annually

In Exercises 17–20, find the annual interest rate r that will yield the given future value starting from the given principal over the specified time period.

17. $P = \$3000$, $F = \$4500$, $t = 4$ years, interest compounded annually
18. $P = \$12,600$, $F = \$20,000$, $t = 8$ years, interest compounded monthly
19. $P = \$65$, $F = \$227$, $t = 19$ years, interest compounded daily
20. $P = \$980$, $F = \$1100$, $t = 9$ months, interest compounded quarterly

In Exercises 21–24, find the annual percentage yield (APY) for the given interest rate compounded as specified.

21. $r = 4.7\%$ compounded monthly
22. $r = 10\%$ compounded quarterly
23. $r = 3.95\%$ compounded daily
24. $r = 6.5\%$ compounded monthly

25. For an account with an annual interest rate of 5%, find the annual percentage yield (APY) when it is compounded
    (a) quarterly,   (b) monthly,   (c) daily.

26. For an account with an annual interest rate of 3%, find the annual percentage yield (APY) when it is compounded
    (a) quarterly,   (b) monthly,   (c) daily.

27. Find the future value of an account after 5 years if $540 is deposited initially and the account earns 8% interest compounded
    (a) annually,   (b) quarterly,   (c) monthly,   (d) daily.

28. Find the future value of an account after 12 years if $3920 is deposited initially and the account earns 6.2% interest compounded
    (a) annually,   (b) quarterly,   (c) monthly,   (d) daily.

29. In January 2012, Ascencia Bank was offering an interest rate of 1.15% compounded monthly on 3-year certificates of deposit, whereas Bank of Internet USA was offering 1.14% compounded daily. If you wanted to invest in a 3-year certificate of deposit, which bank should you choose?

30. A woman wants to set up a college savings account for her granddaughter. If the account earns 5.4% compounded monthly, how much should she invest today so that the account will be worth $50,000 in 18 years?

31. A man has set a goal of saving $15,000 to bankroll the launching of his own business in 2 years. If he has $12,000 to invest today, what interest rate, compounded quarterly, would he have to earn on his investment so that it would grow to $15,000 in 2 years?

32. Suppose you lend $850 to a friend, and your friend agrees to pay the loan back in 3 years at 12% interest compounded quarterly. How much will the friend owe you at the end of 3 years?

33. When Dallas Cowboy linebacker DeMarcus Ware renegotiated his contract in October 2009, he received a $20 million signing bonus, but payment of $1 million of the bonus was deferred to March 2011. Assuming the Cowboys could have made an investment at 4.5% interest compounded monthly, what amount of money would they have needed to invest in October 2009 so that the investment would be worth the $1 million they had to pay DeMarcus Ware in March 2011?

34. A woman who is now 28 years old invests $30,000 in a retirement account. What interest rate, compounded monthly, would she have to earn in order for the account to be worth $500,000 when she retires at age 65?

35. It is quite common for investors to settle for interest rates far below what they could be earning. In January 2012, the United States average interest rate for 5-year certificates of deposit with a minimum deposit of $10,000 was 1.20% compounded daily, whereas one of the best rates in the country, 1.90% compounded daily, was being offered by Intervest National Bank. How much more interest would be earned on a $10,000 5-year certificate if it is invested at 1.90% compounded daily rather than 1.20% compounded daily?

36. California savings banks paid 15% interest in 1860, not too long after the Gold Rush. If the interest had been compounded quarterly, how long would it have taken a $4000 deposit to grow to $10,000?

37. A family has set a goal of saving $18,000 to be used toward the purchase of a new swimming pool. If they put $14,000 into an account earning 7% interest compounded daily, how long will they have to wait until the account has $18,000 in it?

38. Based on the Consumer Price Index, the rate of inflation in the United States during the high inflation years of President Jimmy Carter's term, beginning in January 1977 and ending in January 1981, was about 10.7% per year. What salary in 1977 would be equivalent to a $20,000 salary in 1981?

39. Tuition at the University of Iowa for in-state liberal arts majors grew by about 8.2% per year between 2002 and 2012. In 2012, the annual tuition was $7765. Assuming it will continue to grow at a rate of about 8.2% per year, what will tuition be at the University of Iowa in the year 2025?

40. Suppose the rate of inflation for the next 8 years is predicted to be 3% per year. How much would a house that costs $75,000 today be expected to cost in 8 years?

41. Based on the Consumer Price Index, the rate of inflation in the United States between 1980 and 2011 was about 3.3% per year. What salary in 2011 would be equivalent to a $26,000 salary in 1980?

42. In 1959, the rate for a first-class letter mailed within the United States was 4¢. In 2012, it was 45¢. Find the annual rate of growth (inflation rate) in the cost of mailing a letter from 1959 to 2012.

43. Suppose the rate of inflation for the next 4 years is expected to be 4.5% per year and you plan to save up enough money to purchase a $100,000 home 4 years from now. How much would a home that will cost $100,000 in 4 years be worth today?

44. The inflation rate in Zimbabwe was an amazing 231,000,000% in July 2008. If this rate had continued compounding annually for two more years, how much would an item that cost 5 Zimbabwe dollars in July 2008 have cost two years later?

45. If a deposit is made into an account earning 3.92% compounded daily, how long will it take the investment to double in value?

46. If a deposit is made into an account earning 6% interest compounded monthly, how long will it take the amount in the account to triple in value?

47. Suppose $950 is deposited into an account earning 4% interest compounded daily. After 2 years, all of the money is withdrawn from the account and reinvested in an account earning 9% interest compounded quarterly for 6 more years. How much will the investment be worth after the entire 8 years?

48. Suppose $3000 was deposited into a savings account in a bank. For the first 4 years the bank paid interest at a rate of 3.9% compounded monthly. For the next 2 years the bank paid interest at a rate of 4.2% compounded monthly. How much was in the account at the end of the entire 6 years?

49. One-thousand-year railroad bonds were issued in 1861 when interest rates on railroad bonds averaged 6.33%. How much would a bond purchased for $10,000 in 1861 be worth in 2861 assuming
    (a) simple interest of 6.33% is paid?
    (b) the interest is compounded monthly with an annual rate of 6.33%?

50. A woman deposits $500 into an account earning 4% interest compounded monthly. One year later she deposits an additional $300 into the same account. What will be the total amount in the account 4 years after she made her first deposit?

51. A $9200 deposit is made into an account earning 6.4% interest compounded quarterly. Two years later, an additional $3300 is deposited into the same account. What will be the total amount in the account 6 years after the first deposit was made?

52. In 2012, Discover Bank offered a 5-year certificate of deposit with an annual percentage yield (APY) of 1.80%. The interest was compounded daily. Find the annual interest rate for this certificate of deposit.

53. In 2012, the in-state tuition and fees for 4 years at Florida State University were $22,056. The Florida Prepaid College Program allows tuition and fees to be prepaid. In 2012, one way to prepay the in-state tuition and fees at a 4-year public university for a third-grade student who would enter college 10 years later in 2022 was with a lump sum of $45,613.72.

    (a) If the inflation rate for tuition and fees at Florida State University is 9% per year from 2012 until 10 years later in 2022, what would the tuition and fees be in 2022?

    (b) Assuming the lump-sum payment of $45,613.72 in 2012 would be worth the tuition and fees amount found in part (a) in 2022, what interest rate, compounded annually, would have been earned?

54. In the initial Inter Change Bank judgment discussed in Branching Out 3.3, the investment of $600 million made in 1966 was claimed to be worth $125,444,300,221 in 1994.

    (a) Find the interest rate that would give the same growth if simple interest were paid.

    (b) Find the interest rate that would give the same growth if interest were compounded monthly.

    (c) If the interest had been compounded at 1% interest per week, compounded weekly, find the value of the $600 million investment in 1994.

2.4 THE REWARDS OF SYSTEMATIC SAVINGS

In many cases when we want to save money, we do not have a large amount to deposit at the outset. For instance, when parents try to save for their children's college education, they often do so by setting aside a small amount of money each month to save enough money over a long period of time. We call an account in which a fixed amount of money is deposited at regular intervals a **systematic savings plan.** These kinds of plans are sometimes called annuities, although the word annuity has several different meanings.

To calculate the amount of money that accumulates in a systematic savings plan, we must take into account the fact that because the deposits are made at different times, the amount of interest each deposit earns will be different. Let's see how this situation works by considering a particular example.

A systematic savings plan can finance college tuition.

Suppose $80 is deposited at the end of each month for 5 months into a savings account earning 6% interest compounded monthly. How much money, including interest, will be in the account at the end of 5 months? This is simply the sum of the future values for the five deposits to be made at the end of each month. To figure this out, we look at each deposit separately. The first deposit is made at the end of the first month, so it will have been in the account earning interest for 4 months. Using the compound interest formula with $P = \$80$, $r = 0.06$, $t = \dfrac{4}{12}$ year, and $n = 12$, we find that the first deposit will be worth

$$80\left(1 + \frac{0.06}{12}\right)^{12 \cdot \frac{4}{12}} = 80(1.005)^4.$$

The second deposit will only be in the account for 3 months, so $t = \dfrac{3}{12}$ in the compound interest formula and we see that the second deposit will be worth

$$80\left(1 + \frac{0.06}{12}\right)^{12 \cdot \frac{3}{12}} = 80(1.005)^3.$$

Following this reasoning, we see that the third deposit will be worth

$$80\left(1 + \frac{0.06}{12}\right)^{12 \cdot \frac{2}{12}} = 80(1.005)^2,$$

and the fourth deposit will be worth

$$80\left(1 + \frac{0.06}{12}\right)^{12 \cdot \frac{1}{12}} = 80(1.005)^1.$$

The fifth deposit is made at the very end of the 5 months, so it will have earned no interest and is only worth $80. Adding up the future values of each of the separate deposits, we see that at the end of 5 months the entire account will be worth

$$80(1.005)^4 + 80(1.005)^3 + 80(1.005)^2 + 80(1.005) + 80 \approx \$404.02.$$

Finding the amount in the account this way was not too bad for 5 months, but the work required to find the amount in the account after several years looks overwhelming. This is where a little mathematics goes a long way in simplifying the calculations.

We first notice that the reasoning we used to come up with the sum for the entire amount in the account would work for any number of months. If the deposits are made at the end of each month for k months, then the amount in the account after k months is the sum of the future values:

$$80(1.005)^{k-1} + 80(1.005)^{k-2} + \cdots + 80(1.005) + 80$$
$$= 80\left[(1.005)^{k-1} + (1.005)^{k-2} + \cdots + (1.005) + 1\right].$$

Now we can apply the formula for the sum of a geometric series. This formula says that for any number a not equal to 1 we have

$$a^{k-1} + a^{k-2} + \cdots + a + 1 = \frac{a^k - 1}{a - 1}.$$

Applying this formula to the last equation, we find that the amount in the account after k months is given by

$$80\left(\frac{(1.005)^k - 1}{1.005 - 1}\right) = 80\left(\frac{(1.005)^k - 1}{0.005}\right).$$

This formula is easy to use even if the number of months k is very large.

A CLOSER *look*

ADDING IT UP: THE SUM OF A GEOMETRIC SERIES

The formula for the sum of a geometric series

$$a^{k-1} + a^{k-2} + \cdots + a + 1 = \frac{a^k - 1}{a - 1},$$

for $a \neq 1$, follows from the identity

$$(a-1)(a^{k-1} + a^{k-2} + \cdots + a + 1) = a^k - 1.$$

This last equation can be seen to be true by multiplying out the left-hand side of the equation and seeing that everything cancels out except $a^k - 1$. For instance, if $k = 4$ we have

$$(a-1)(a^3 + a^2 + a + 1) = (a^4 + a^3 + a^2 + a) - (a^3 + a^2 + a + 1) = a^4 - 1.$$

Now let's take these ideas to come up with a formula for the general situation. We notice that the 80 is just the amount of each deposit, the number 0.005 is the periodic interest $\frac{r}{n}$, where r is the annual interest rate and n is the number of periods each year. Also, the number k is the number of periods the account is held, so $k = nt$, where t is the number of years the account is held. Putting all of this together, for nt an integer, we have the following general formula for the future value of a systematic savings plan.

Systematic Savings Plan Formula

$$F = D\left(\frac{\left(1 + \frac{r}{n}\right)^{nt} - 1}{\frac{r}{n}}\right) \quad \text{where} \quad \begin{cases} F = \text{future value after } t \text{ years} \\ D = \text{amount of each deposit} \\ r = \text{annual interest rate} \\ n = \text{number of periods per year} \\ t = \text{time (in years)} \end{cases}$$

Deposits are made at the end of each period.

It is important to note that the intervals at which the deposits are made have to be the same as the intervals at which interest is compounded in order for the formula to apply. It is not very difficult to adapt the formula to the situation when this is not the case, but we will not pursue it here, and instead leave the derivation of the formula as an exercise.

In deriving the systematic savings plan formula, we also assumed that the interest rate did not vary and that the deposits were all the same size and at regular intervals. If either of these were not the case, a simple formula like the one we found would not be possible. In most true savings situations, interest rates vary over time and the deposits vary in size and may be irregularly spaced. However, we can still use our systematic savings plan formula to make reasonable estimates. Such estimates are key to long-term financial planning.

EXAMPLE 1 *Finding the Future Value*

Suppose you deposit $200 at the end of each month into an account earning 5.4% interest compounded monthly. How much will be in the account, how much of the future value will be from deposits, and how much of the future value will be from interest after making deposits for

(a) 15 years?
(b) 40 years?

SOLUTION:

(a) The amount in the account after 15 years of deposits is the future value F given by the systematic savings formula with $D = 200$, $r = 0.054$, $t = 15$, and $n = 12$. Substituting into the formula, we find that

$$F = 200\left(\frac{\left(1 + \dfrac{0.054}{12}\right)^{12 \cdot 15} - 1}{\dfrac{0.054}{12}}\right)$$

$$F = 200\left(\frac{(1.0045)^{180} - 1}{0.0045}\right)$$

$$F \approx 200(276.40602937)$$

$$F \approx \$55{,}281.21.$$

To find the total amount deposited, we simply multiply the amount of each deposit, $200, by the number of deposits $nt = 12 \cdot 15 = 180$. We find that the total amount deposited is

$$\$200 \cdot 180 = \$36{,}000.$$

Because only $36,000 of the future value of $55,281.01 came from deposits, the rest must have come from interest. So the total amount of interest accumulated is given by

$$\$55{,}281.21 - \$36{,}000 = \$19{,}281.21.$$

(b)   The amount F in the account after 40 years is given by

$$F = 200\left(\frac{\left(1 + \dfrac{0.054}{12}\right)^{12 \cdot 40} - 1}{\dfrac{0.054}{12}}\right)$$

$$F = 200\left(\frac{(1.0045)^{480} - 1}{0.0045}\right)$$

$$F \approx 200(1695.38301294)$$

$$F \approx \$339{,}076.60.$$

Because $12 \cdot 40 = 480$ deposits of $200 each are made over 40 years, the amount of the future value from deposits is

$$\$200 \cdot 480 = \$96{,}000,$$

and the amount of the future value from interest is given by

$$\$339{,}076.60 - \$96{,}000 = \$243{,}076.60.$$

After 40 years, more of the money in the account is from interest than from deposits. In fact, as we can see in the graph in Figure 2.2, after only about 23 years, the interest accumulated in the account had surpassed the amount deposited. If the account were continued even longer, an even higher percentage of the account would come from interest.

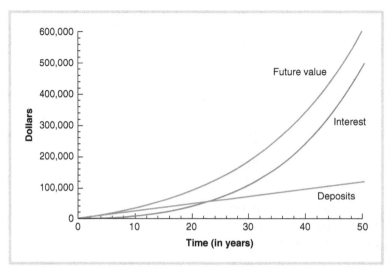

FIGURE 2.2 Comparison of the accumulated interest and deposits in a systematic savings account.

EXAMPLE 2 *Saving for College*

The parents of a kindergarten student want to save $200,000 to pay for the estimated cost of her college education. They plan to invest their savings in a mutual fund, and based on the past performance of the fund and its projected future performance, they estimate that it will give returns that are equivalent to earning about 8% interest compounded quarterly. How much should the parents deposit at the end of each quarter into an account paying 8% interest compounded quarterly so that the account will be worth $200,000 in 13 years?

SOLUTION:

In this case we have $F = 200{,}000$, $r = 0.08$, $t = 13$, $n = 4$, and we want to find the amount of each deposit D. Substituting into the systematic savings plan formula and solving for D, we get

$$200{,}000 = D\left(\frac{\left(1 + \frac{0.08}{4}\right)^{4 \cdot 13} - 1}{\frac{0.08}{4}}\right)$$

$$200{,}000 = D\left(\frac{(1.02)^{52} - 1}{0.02}\right)$$

$$200{,}000 \approx D(90.01640927)$$

$$D \approx \frac{200{,}000}{90.01640927}$$

$$D \approx \$2221.82.$$

Making a deposit of $2221.82 each quarter means that the parents will have to budget about $740 each month toward the fund—a sobering thought.

branching OUT

THE TAX ADVANTAGES OF IRAs

Image © Alexander Raths, 2012. Shutterstock, Inc.

Individual retirement accounts (IRAs) are investments with tax advantages for the investor. One kind of IRA, known as a traditional IRA, does not require the payment of taxes on interest earned until the money is withdrawn after retirement. In addition, for an investor who is not covered by an employer-sponsored retirement plan or whose income is below a specified level, the money invested in a traditional IRA is deductible from his or her income. In this case, all the money in the account is taxable on withdrawal. Individuals under 50 years of age can invest up to $5000 each year in a traditional IRA.

To see how the tax advantages affect the growth of an investment, consider the example of an individual in the 28% tax bracket who can take full advantage of a traditional IRA's tax breaks. Suppose the person invests $5000 at the end of each year for 30 years in a traditional IRA at 7% interest, compounded annually. The amount in the account after t years is given by

$$F = 5000\left(\frac{(1.07)^t - 1}{0.07}\right).$$

However, we assume that the withdrawals will be taxed at 28% on withdrawal, so the true value of the account to the investor after t years is only $100\% - 28\% = 72\%$ of the total or

$$F = (0.72)\left(5000\left(\frac{(1.07)^t - 1}{0.07}\right)\right).$$

By comparison, if the individual invested in a regular account, then the $5000 would first be taxed at 28%, leaving only 72% of $5000, or $(0.72) \times \$5000 = \$3600$ to be deposited each year. In addition, because the interest will be taxed at 28% each year, we can consider the true interest rate to be $(0.72) \times 7\% = 5.04\%$.

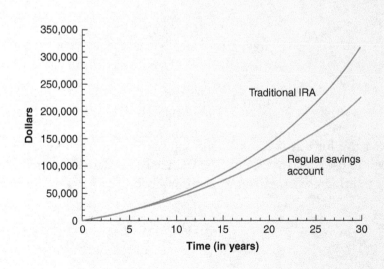

Therefore, after taxes, the amount in the account after t years is given by

$$F = 3600 \left(\frac{(1.0504)^t - 1}{0.0504} \right).$$

The withdrawals are not taxable, so this is the true value of the account for the investor after t years. The true value of the accounts to the investor over the 30-year period is shown in the graph.

After 30 years, the traditional IRA account is worth $340,058.83 and the account with no tax breaks is worth only $240,829.27. The tax advantages are even greater if the investor is able to avoid Social Security taxes on the amount deposited, as is possible in some cases, or if the investor is in a lower tax bracket after retirement.

Most investors have the option of choosing another kind of IRA, called a Roth IRA. With this kind of IRA, the money invested cannot be deducted but the interest earnings are never taxed. Which kind of IRA is a better option for an investor depends on the investor's tax bracket before and after retirement and on how long the money is invested.

EXAMPLE 3 *Finding Time*

Suppose you want to have $3000 available for the down payment on a car. If you deposit $120 every month, at the month's end, into an account earning 6% compounded monthly, how long will it take until you have the down payment?

SOLUTION:

In this situation we know $F = 3000$, $D = 120$, $r = 0.06$, and $n = 12$. We use the systematic savings plan formula to solve for the number of years t it will take to save $3000. Substituting into the formula, we have

$$3000 = 120 \left(\frac{\left(1 + \frac{0.06}{12}\right)^{12t} - 1}{\frac{0.06}{12}} \right)$$

$$3000 = 120 \left(\frac{(1.005)^{12t} - 1}{0.005} \right).$$

How do we solve for this t? A systematic approach is the key. We know that in the end we need to use logarithms, but first we need to isolate the term $(1.005)^{12t}$. We use algebra to bring us step by step to our goal.

$$\frac{3000}{120} = \frac{(1.005)^{12t} - 1}{0.005}$$

$$25 = \frac{(1.005)^{12t} - 1}{0.005}$$

$$(25)(0.005) = (1.005)^{12t} - 1$$

$$0.125 = (1.005)^{12t} - 1$$

$$0.125 + 1 = (1.005)^{12t}$$

$$1.125 = (1.005)^{12t}$$

Now the term $(1.005)^{12t}$ is isolated, so we take logarithms of both sides of the equation and solve for t.

$$\log(1.125) = \log(1.005)^{12t}$$

$$\log(1.125) = 12t \log(1.005)$$

$$t = \frac{\log(1.125)}{12 \log(1.005)}$$

$$t \approx 1.97 \text{ years.} \qquad \blacksquare$$

The time we solved for in Example 3 is not exactly the end of a period. The next period will end at 2 years or 24 months, so in fact when $t = 1.97$, because the last deposit has not been made and interest has probably not been paid for the last period, the account will not be worth $3000. In fact, it will never be worth exactly $3000, and its value will exceed $3000 when $t = 2$. As with solving for time in the case of compound interest problems, we will not be concerned with this here. So, for instance, in Example 3, we will keep the answer $t = 1.97$. Just keep in mind that the times that we solve for are approximate. They should be sufficiently accurate for financial planning.

E X E R C I S E S FOR SECTION 2.4

In Exercises 1–4, find the future value F for a systematic savings plan with the given deposits made at the ends of the compounding periods for the time period and interest rate specified.

1. $D = \$640$, $t = 15$ years, $r = 5.6\%$ compounded quarterly
2. $D = \$25$, $t = 30$ years, $r = 4.8\%$ compounded monthly
3. $D = \$700$, $t = 10$ years, $r = 11\%$ compounded monthly
4. $D = \$35$, $t = 50$ years, $r = 9.2\%$ compounded quarterly

In Exercises 5–8, find the amount of each deposit D for a systematic savings plan with the deposits made at the ends of the compounding periods given the future value, time period, and interest rate specified.

5. $F = \$40,000$, $t = 11$ years, $r = 5\%$ compounded quarterly
6. $F = \$892$, $t = 3$ years, $r = 4\%$ compounded quarterly
7. $F = \$5000$, $t = 21$ months, $r = 8\%$ compounded monthly
8. $F = \$3,000,000$, $t = 40$ years, $r = 6.2\%$ compounded monthly

In Exercises 9–12, find the time t, in years, for a systematic savings plan with the given deposits made at the ends of the compounding periods to reach the given future value under the specified interest rate.

9. $D = \$75$, $F = \$1200$, $r = 6\%$ compounded monthly
10. $D = \$800$, $F = \$25,000$, $r = 5.8\%$ compounded monthly
11. $D = \$130$, $F = \$4000$, $r = 12\%$ compounded quarterly
12. $D = \$680$, $F = \$200,000$, $r = 9.3\%$ compounded quarterly

13. A woman has \$75 deducted from her paycheck at the end of each month and put into a savings account earning 9% interest compounded monthly. She continues these deposits for 10 years.
    (a) How much will the account be worth after 10 years?
    (b) How much of the future value will be from deposits?
    (c) How much of the future value will be from interest?

14. Suppose a 27-year-old man bought an annuity (a systematic savings plan) through the Jackson National Life Insurance Company in 1997. The plan calls for deposits of \$100 at the end of each month and pays an interest rate of 6.25% compounded monthly.
    (a) How much will the annuity be worth when the man is 65 years old in 2035?
    (b) How much of the future value will be from deposits?
    (c) How much of the future value will be from interest?

15. A family wants to save \$7000 to use toward the purchase of a piano. If they deposit \$600 at the end of each quarter into an account earning 9% interest compounded quarterly, how long will it take for them to save \$7000?

16. Suppose you would like to have \$24,000 available to be used toward the down payment on a house in 3 years. If you can invest the money in an account earning 6.4% interest compounded quarterly, how much should you deposit at each quarter's end so that you will have \$24,000 in 3 years?

17. Suppose a man makes monthly deposits into a retirement savings account earning 7% compounded monthly. How much will he have in the account when he retires at age 70 if
    (a) he makes deposits of \$100 at each month's end beginning at age 22?
    (b) he makes deposits of \$200 at each month's end beginning at age 40?

18. From 1928 to 2011, U.S. stocks posted an average annual return of 11.20%. During the same time, long-term U.S. treasury bonds gave an average annual return of 5.41%, and U.S. treasury bills earned 3.66% per year on average. Making the simplifying assumption that the interest rates were constant, for each of the three investments, find the amount you would have in 2011 if you had made annual investments of \$1000 at the end of each year from 1928 to 2011.

19. A couple predicts that they will need to save \$250,000 for their child's college education, and they predict that they will be able to earn about 9% interest, compounded monthly, on their investments.
    (a) If they begin the deposits at the end of each month when their child is a new-born, so that they have 18 years of deposits, how large must each deposit be?
    (b) If they do not begin making deposits until the child is 10 years old, so that they have only 8 years of deposits, how large must each deposit be?

20. On January 24, 2012, the United States average interest rate for money market accounts was 0.51%, whereas one of the best rates in the country, 0.90%, was being offered by First Internet Bank of Indiana. If $100 is invested at the end of each month into an account, how long will it take the account to grow to $10,000 if it is invested at
    (a) 0.51% compounded monthly?
    (b) 0.90% compounded monthly?

21. The Mississippi Prepaid Affordable College Tuition Program (MPACT) is a prepaid tuition program through which the college tuition and fees at any Mississippi public university or college may be paid for in advance. In 2011, to cover the costs of in-state tuition at a 4-year university for a newborn who will enter college 18 years later, purchasers of the plan had the option of paying one lump-sum payment of $21,043 immediately, payments of $420 at each month's end for 5 years beginning immediately, or payments of $182 at each month's end for 17 years and 4 months. Show that the future values after 17 years and 4 months of each of these deposits into an account with an interest rate of 7.8% compounded monthly are fairly close in value.

22. Suppose $600 is deposited at the end of each month into an account earning 10% interest compounded monthly. After 8 years the deposits are discontinued, but the money is left in the account and continues to earn the same interest. How much will be in the account 15 years after it was first opened?

23. A woman pays $130 at each quarter's end into an account earning 5.4% interest compounded quarterly. After 10 years of deposits, she changes the amount of the deposits to $200 per quarter. If the money continues to earn interest at the same rate, how much will be in the account 30 years after it was first opened?

24. A family currently has $10,000 of the $20,000 they want to have available for a down payment on a house. They invest this $10,000 in a savings account paying 5.7% interest compounded monthly, and at the end of each month they deposit another $150 into the account. How long will it take them to save enough for the down payment?

25. It would not be unusual to have a systematic savings plan in which interest is compounded more often than deposits are made. For instance, annual deposits might be made into an account with interest compounded monthly. The systematic savings formula derived in the text does not directly apply to this case, but it can be adapted.
    (a) Find a formula for the future value F after t years of a systematic savings plan with an annual interest rate r, compounded n times a year, into which deposits of size D are made at the end of every k periods. (Assume that nt is exactly divisible by k.)
    (b) Use the formula from part (a) to find the future value after 7 years of an account with interest at 6% compounded monthly into which deposits of $50 are made at the end of each quarter.

26. Although it is not possible to algebraically solve exactly for the interest rate r in the systematic savings plan formula, r can be estimated by using a calculator, spreadsheet, or computer program that will perform such estimates or by making a series of guesses for r. When using the guessing method, you should determine if your first guess was too high or too low and adjust accordingly,

continuing to adjust your guesses until the value of r you have found gives a reasonably close answer. Suppose deposits of $230 are made at the end of each month into an account earning interest compounded monthly. After 6 years, the account is worth $20,432. Use some method to estimate the annual interest rate r to two decimal places.

27. Derive a general formula for the future value of a systematic savings plan when deposits are made at the beginning, rather than the end, of each period.

2.5 AMORTIZED LOANS

Most of us borrow money many times over the course of our lifetime. We may take loans to pay tuition in college, purchase a car, start a business, or buy a home. Loans are contracts between a lender and a borrower who agrees to pay back the loan under some agreed upon terms. With an **amortized loan**, a lender loans a borrower a lump sum of money, and the borrower pays back the loan by making *equal* payments at some regular intervals. To earn money, the lender charges interest on the balance of the loan. The payments are due at the end of each compounding period, and each payment covers the periodic interest due on the current balance of the loan. In addition, each payment includes some extra money that is used to bring down the balance so that the loan will eventually be paid off. In the early payments, the portion of the loan that is applied to paying the interest is higher than later in the life of the loan when the balance has been reduced. The benefits of setting up a loan with regular equal payments is that it gives the borrower a definite schedule for paying off the loan, it assures the lender of a flow of cash, and it provides a regular check as to whether the borrower is meeting his or her obligations.

When lenders loan money at an interest rate r, they expect the income stream coming in from the payments to be equivalent in value to investing the principal themselves at the same interest rate r. In other words, the future value of the series of payments must be the same as the future value of investing the principal in an account with the same rate and compounding period. We will now see how this idea can be used to find a formula that allows us to figure out exactly what size the equal payments must be in order for an amortized loan to be paid off in a specified period of time.

The initial amount of money borrowed is called the **principal** or **present value** of the loan, and we denote it by the letter P. Suppose the interest rate on the loan is r, compounded

A house purchase is usually paid with an amortized loan.

n times a year. Then the future value F of the initial loan amount P after t years is given by the compound interest formula

$$F = P\left(1 + \frac{r}{n}\right)^{nt}.$$

If the loan is paid off in t years with payments of R dollars at the end of each period, then the future value F of this series of payments is given by the systematic savings plan formula

$$F = R\left(\frac{\left(1 + \frac{r}{n}\right)^{nt} - 1}{\frac{r}{n}}\right).$$

To find a formula that relates the payment amount R to the principal, rate, compounding period, and length of the loan, we use the key fact that these two future values must be equal. Setting them equal to each other, we have

$$P\left(1 + \frac{r}{n}\right)^{nt} = R\left(\frac{\left(1 + \frac{r}{n}\right)^{nt} - 1}{\frac{r}{n}}\right)$$

We can now solve for P as follows:

$$P = \frac{1}{\left(1 + \frac{r}{n}\right)^{nt}}\left[R\left(\frac{\left(1 + \frac{r}{n}\right)^{nt} - 1}{\frac{r}{n}}\right)\right]$$

$$= R\left(\frac{1 - \left(1 + \frac{r}{n}\right)^{-nt}}{\frac{r}{n}}\right).$$

Thus, for nt an integer, we have the following loan formula.

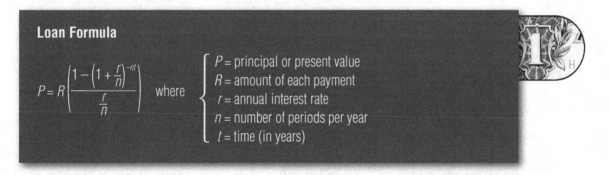

Loan Formula

$$P = R\left(\frac{1 - \left(1 + \frac{r}{n}\right)^{-nt}}{\frac{r}{n}}\right) \quad \text{where} \quad \begin{cases} P = \text{principal or present value} \\ R = \text{amount of each payment} \\ r = \text{annual interest rate} \\ n = \text{number of periods per year} \\ t = \text{time (in years)} \end{cases}$$

The preceding derivation illustrates a general mathematical technique. When the same quantity can be computed in two different ways, by setting the two expressions equal a potentially new formula or identity often may be discovered.

EXAMPLE 1 *Cost of Buying a Car*

Image © egd, 2012. Shutterstock, Inc.

Suppose you want to purchase a car and the added charges for taxes, title, license, and fees bring the total to \$16,032.31. You must make a 10% down payment and pay for the remainder through a car loan at an interest rate of 7.9% compounded monthly. You are to repay the loan with monthly payments for 4 years.

(a)   What are the monthly payments?
(b)   What is the total amount you make in payments over the life of the loan?
(c)   How much interest will you pay over the life of the loan?

SOLUTION:

(a)   To find the monthly payments, we must first find the principal for the loan that will be the purchase price minus the down payment. The down payment is 10% of $16,032.31 or

$$(0.10)(16,032.31) = 1603.231 \approx \$1603.23,$$

so the principal for the loan will be given by

$$P = 16,032.31 - 1603.23 = \$14,429.08.$$

Now, to find the monthly payments we substitute $P = 14,429.08$, $r = 0.079$, $n = 12$, and $t = 4$ into the loan formula and solve for the monthly payment R:

$$14,429.08 = R\left(\frac{1 - \left(1 + \frac{0.079}{12}\right)^{-12 \cdot 4}}{\frac{0.079}{12}}\right)$$

$$14,429.08 \approx R\left(\frac{1 - (1.00658333)^{-48}}{0.00658333}\right)$$

$$14,429.08 \approx R(41.04078199)$$

$$R \approx \frac{14,429.08}{41.04078199}$$

$$R \approx \$351.58.$$

(b)   You will make $nt = 12 \cdot 4 = 48$ payments. Because the amount of each payment is $R = \$351.58$, the total amount paid over the life of the loan is

$$\$351.58 \cdot 48 = \$16,875.84.$$

(c)   The interest you will pay over the life of the loan is the difference between the total amount you paid and the initial principal P. Thus the total interest paid is

$$\$16,875.84 - \$14,429.08 = \$2446.76. \qquad \blacksquare$$

The total interest paid over the life of the loan is sometimes called the *finance charge*. The term finance charge may also refer to the monthly interest charged on a revolving credit account such as a credit card account.

Sometimes borrowers know the maximum periodic payments they can afford and they want to know how much money they can borrow. We examine this question in the following example.

EXAMPLE 2 *Student Loan*

A college senior does not have enough money to cover expenses and plans to get extra money through working and through a student loan that will have quarterly payments for 6 years. She estimates that after graduation she can afford to make quarterly loan payments of $300. If the loan is made at an interest rate of 7% compounded quarterly, how much can she afford to borrow?

SOLUTION:

In this situation we know the quarterly payments will be $R = 300$, and we know $r = 0.07$, $n = 4$, and $t = 6$. Substituting into the loan formula, we find the principal P:

$$P = 300\left(\frac{\left(1 - \left(1 + \frac{0.07}{4}\right)^{-4 \cdot 6}\right)}{\frac{0.07}{4}}\right)$$

$$P = 300\left(\frac{1 - (1.0175)^{-24}}{0.0175}\right)$$

$$P \approx 300(19.46068565)$$

$$P \approx \$5838.21.$$

We see that she can afford to borrow $5838.21. ∎

With most loans, the borrower has the option of paying more than the required payments each period. By doing so, not only is the loan paid off more quickly, but also the total interest paid is reduced, and therefore the total amount paid over the life of the loan is reduced. The savings achieved by such extra payments can be quite striking. We see how this works in the next example.

EXAMPLE 3 *Saving by Paying Off a Loan Quickly*

A couple purchased a house by taking out a $125,000 mortgage (loan) with monthly payments for 30 years. The interest rate on the loan was 8.25% compounded monthly.

(a) How much would the scheduled payments on the loan be?
(b) If they pay $100 extra each month, how long will it take to pay off the loan?
(c) How much does the couple save in payments over the life of the loan by paying $100 extra each month?

SOLUTION:

(a) We have $P = 125{,}000$, $r = 0.0825$, $n = 12$, and $t = 30$. Using the loan formula to solve for the scheduled payment R, we get

$$125{,}000 = R\left(\frac{1 - \left(1 + \frac{0.0825}{12}\right)^{-12 \cdot 30}}{\frac{0.0825}{12}}\right)$$

$$125{,}000 = R\left(\frac{1 - (1.006875)^{-360}}{0.006875}\right)$$

$$125{,}000 \approx R(133.10853891)$$

$$R \approx \frac{125{,}000}{133.10853891} \approx \$939.08.$$

(b) By paying \$100 extra each month, the payment is now

$$R = \$939.08 + \$100 = \$1039.08.$$

To find how long it will take to repay the loan with this new payment amount, we substitute $P = 125{,}000$, $R = 1039.08$, $r = 0.0825$, and $n = 12$ into the loan formula, and solve for t.

$$125{,}000 = 1039.08\left(\frac{1 - \left(1 + \frac{0.0825}{12}\right)^{-12t}}{\frac{0.0825}{12}}\right)$$

$$125{,}000 = 1039.08\left(\frac{1 - (1.006875)^{-12t}}{0.006875}\right).$$

To solve for t, we need to use logarithms, but first we need to isolate the term $(1.006875)^{-12t}$.

$$\frac{125{,}000}{1039.08} = \frac{1 - (1.006875)^{-12t}}{0.006875}$$

$$120.2987258 \approx \frac{1 - (1.006875)^{-12t}}{0.006875}$$

$$(120.2987258)(0.006875) \approx 1 - (1.006875)^{-12t}$$

$$0.82705374 \approx 1 - (1.006875)^{-12t}.$$

Next, we subtract 1 from each side of the equation:

$$0.82705374 - 1 \approx -(1.006875)^{-12t}$$

$$-0.17294626 \approx -(1.006875)^{-12t}.$$

We now multiply both sides of the equation by −1 and get

$$(-1)(-0.17294626) \approx (-1)(-(1.006875)^{-12t})$$

$$0.17294626 \approx (1.006875)^{-12t}.$$

Now we take the logarithms of both sides of the equation and solve for t:

$$\log(0.17294626) \approx \log(1.006875)^{-12t}$$

$$\log(0.17294626) \approx -12t \log(1.006875)$$

$$t \approx \frac{\log(0.17294626)}{-12 \log(1.006875)}$$

$$t \approx 21.34 \text{ years.}$$

We see that by paying an extra $100 each month, the loan is paid off in 21.34 years rather than the original 30 years. As with solving for time in the previous sections, we must keep in mind that the answer is approximate.

(c) The total amount paid over the life of the loan is the amount of each payment R multiplied by the number of payments nt. For the scheduled payments, this total is given by

$$(\$939.08)(12)(30) = \$338,068.80.$$

When paying $100 extra each month, the payment R is $1039.08 and the time t is roughly 21.34 years, so the total amount paid over the life of the loan is approximately

$$(\$1039.08)(12)(21.34) = \$266,087.61.$$

We see that the total amount the couple saves in interest over the life of the loan by paying $100 extra each month is approximately

$$\$338,068.80 - \$266,087.61 = \$71,981.19. \quad \blacksquare$$

It is important to note that in Example 3 we did not consider interest that could be earned by investing the $100 each month. Furthermore, we ignored taxes on earned interest and tax deductions on mortgages. All these factors should be taken into account when making the decision of whether or not to pay off a loan more quickly.

Finding a Loan Balance

When you have a loan, you may want to know how much you still owe on the loan at a given point in time. This amount is called the **loan balance**. The loan balance is the present value of the series of payments that are left to make. In other words, the loan balance is the principal that would have to be borrowed today at the specified rate to yield the series of payments remaining to be paid. To compute the balance of a loan, we use the loan formula in a different way. If we let R be the usual payment amount and t be the amount of time left in the life of the loan, then the present value P given by the loan formula is the current balance of the loan. We now see how this works.

branching OUT

THE MYSTERIOUS APR

Annual percentage rate, or APR, is to loans as annual percentage yield is to savings. However, because of up-front fees that vary from loan to loan, APR is much harder to calculate. In fact, different banks may compute APR slightly differently on exactly the same loan. Federal laws mandate that lenders provide the annual percentage rate for consumer protection. Specifically, finance charge and annual percentage rate are defined in Sections 1605 and 1606 of Chapter 41, Consumer Credit Protection, of Title 15, Commerce and Trade, of the United States Code. The APR must include costs arising as part of the extension of credit such as application fees, loan fees, or points. The APR does not include charges that would arise in a cash transaction such as title fees, deed preparation, or escrow for taxes and insurance.

Let's illustrate with an example. Suppose you finance the purchase of a home through a $100,000 mortgage with 30 years of monthly payments at 9% interest. From the amortization formula, we find the monthly payment R satisfies

$$100{,}000 = R\left(\frac{1 - \left(1 + \frac{0.09}{12}\right)^{-30 \cdot 12}}{\frac{0.09}{12}}\right)$$

Solving for R in the usual way, we find $R = \$804.62$. Suppose the loan requires a $100 application

fee and an up-front payment of 1 point. Then you pay the bank the $100 application fee and 1% of $100,000, or $1000, for a total of $1100. Thus, the bank only has to loan you $100,000 − $1100 = $98,900 from its own funds. The interest rate on a loan of $98,900 with monthly payments of $804.62 has an effective interest rate satisfying

$$98{,}900 = 804.62\left(\frac{1 - \left(1 + \frac{r}{12}\right)^{-30 \cdot 12}}{\frac{r}{12}}\right)$$

We cannot combine terms algebraically in any way that allows us to solve exactly for r. However, using a numerical equation solver available on many calculators, spreadsheet programs, or similar computer software packages, we estimate $r \approx 9.124\%$. The closing statement from the bank is required to state this as the APR.

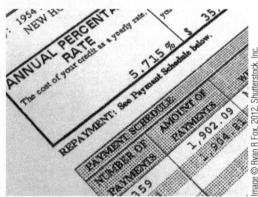

EXAMPLE 4 *Loan Balance*

Suppose a $12,000 loan is taken out at an interest rate of 11.3% compounded monthly with monthly payments for 5 years. After 2 years of payments, what is the balance of the loan?

First we must figure out the payment amount R. We use the loan formula with $P = 12,000$, $r = 0.113$, $n = 12$, and $t = 5$ and solve for R.

$$12,000 = R \left(\frac{1 - \left(1 + \frac{0.113}{12}\right)^{-12 \cdot 5}}{\frac{0.113}{12}} \right)$$

$$12,000 \approx R \left(\frac{1 - (1.00941667)^{-60}}{0.00941667} \right)$$

$$12,000 \approx R(45.67808441)$$

$$R \approx \frac{12,000}{45.67808441}$$

$$R \approx \$262.71.$$

After 2 years of payments, the amount of time left in the life of the loan is 3 years. To find the balance left after 2 years, we use the loan formula with $R = 262.71$, $r = 0.113$, $n = 12$, and $t = 3$. This gives the present value, P, of the remaining series of payments:

$$P = 262.71 \left(\frac{1 - \left(1 + \frac{0.113}{12}\right)^{-12 \cdot 3}}{\frac{0.113}{12}} \right)$$

$$P = 262.71 \left(\frac{1 - (1.00941667)^{-36}}{0.00941667} \right)$$

$$P \approx \$7989.73.$$

Thus the balance of the account after two years is \$7989.73. ■

Amortization Schedules

An **amortization schedule** is a list of payments to be made on a loan that breaks down each payment into principal and interest. Many lenders give borrowers an amortization schedule each year, often in the form of a coupon book. An amortization schedule provides a good record of the total amount of interest paid each year on a loan. These records may be important for tax purposes if the interest is tax deductible, as is the case for home loans and home equity loans. Amortization schedules sometimes include the balance of the loan after each payment, thus providing an easy way to look up the balance of the loan at any time.

In the next example, we see how an amortization schedule is constructed.

EXAMPLE 5 *Amortization Schedule*

Suppose a $2758.35 furniture purchase is financed with a loan at 9.6% interest compounded monthly with monthly payments for 4 months. Write an amortization schedule for the loan.

SOLUTION:

We first use the loan formula to find the amount of each payment. Substituting $P = 2758.35$, $r = 0.096$, $n = 12$, and $t = 4/12$ into the loan formula, we get

$$2758.35 = R\left(\frac{1 - \left(1 + \frac{0.096}{12}\right)^{-12 \cdot (4/12)}}{\frac{0.096}{12}}\right)$$

$$2758.35 = R\left(\frac{1 - (1.008)^{-4}}{0.008}\right)$$

$$2758.35 \approx R(3.92126231)$$

$$R \approx \frac{2758.35}{3.92126231}$$

$$R \approx \$703.43.$$

Now we are ready to set up an amortization schedule. The schedule lists the payment number, the amount of each payment, the amount of interest and principal in each payment, and the loan balance. We can arrange this information in a table that starts out as follows:

Payment Number	Payment	Interest Paid	Principal Paid	Balance
				2758.35
1				
2				
3				
4				

To fill out the next row in the table, we enter $703.43 in the payment column. To find what portion of this payment is interest, we multiply the balance appearing in the previous row by the interest rate for one period $r/n = 0.096/12 = 0.008$. We find

$$\text{interest paid} = (\$2758.35)(0.008) \approx \$22.07.$$

The remainder of the first payment goes toward paying the principal, so we have

$$\text{principal paid} = \$703.43 - \$22.07 = \$681.36.$$

We compute the new balance by subtracting the principal paid from the previous balance:

$$\text{balance} = \$2758.35 - \$681.36 = \$2076.99.$$

Entering this information into the table, we have

Payment Number	Payment	Interest Paid	Principal Paid	Balance
				2758.35
1	703.43	22.07	681.36	2076.99
2				
3				
4				

To fill in the next row, we first enter the payment and then compute the interest paid. It is given by the new balance multiplied by the monthly interest rate as follows:

$$\text{interest paid} = (\$2076.99)(0.008) \approx \$16.62.$$

The principal paid is then given by

$$\text{principal paid} = \$703.43 - \$16.62 = \$686.81,$$

and the new balance is given by

$$\text{balance} = \$2076.99 - \$686.81 = \$1390.18.$$

Entering this information into our amortization schedule we get

Payment Number	Payment	Interest Paid	Principal Paid	Balance
				2758.35
1	703.43	22.07	681.36	2076.99
2	703.43	16.62	686.81	1390.18
3				
4				

Continuing in this way, we complete the next row of the amortization schedule and we have

Payment Number	Payment	Interest Paid	Principal Paid	Balance
				2758.35
1	703.43	22.07	681.36	2076.99
2	703.43	16.62	686.81	1390.18
3	703.43	11.12	692.31	697.87
4				

The last line of an amortization schedule is computed in a somewhat different way, and the last payment may be slightly different from the other payments. In the last line, the interest paid is computed in the usual way and we find that

$$\text{interest paid} = (\$697.87)(0.008) \approx \$5.58.$$

The principal paid in the last payment must pay off the balance of the loan, and therefore we simply let the principal paid equal the previous balance. Finally, we compute the amount of the last payment by adding the interest paid and the principal paid and we find that

$$\text{payment} = \$5.58 + \$697.87 = \$703.45.$$

We see that in this case the last payment was $0.02 higher than the other payments. In an amortization schedule, the last payment may be higher, lower, or the same as the other payments. Completing our amortization schedule, we have

Payment Number	Payment	Interest Paid	Principal Paid	Balance
				2758.35
1	703.43	22.07	681.36	2076.99
2	703.43	16.62	686.81	1390.18
3	703.43	11.12	692.31	697.87
4	703.45	5.58	697.87	0

■

Notice that with each successive payment, the amount of interest paid goes down and the amount of principal paid goes up. This phenomenon always occurs with amortized loans. With a typical 30-year home loan the payments for the first several years consist mostly of interest payments. Because this is the case, one simple way to estimate the size of the payments on a 30-year loan is to compute the monthly interest due on the original balance. The actual payments will be only somewhat higher. This idea will not apply to shorter term loans as you can see in Example 5, where the initial interest is $22.07, but the payments are $703.43 except for the final one.

branching OUT

LOTTERY WINNINGS: LUMP SUM VS. ANNUAL PAYMENTS

In 2012, the New Jersey Lottery gave people playing the game Pick-6 Lotto the option of having any winnings paid immediately in a one-time lump sum or paid in 26 equal annual payments with one payment made immediately and the remaining payments made at the end of each year for the next 25 years. For the February 6, 2012 lottery jackpot, the cash option was a single payment of $3.1 million. For the annual payment option, the winner would receive $4.2 million in 26 equal payments of about $161,538 each. Why was the lump-sum payment so much less than the amount that would be received under the annual payment option?

In this case, the lump sum of $3.1 million is the actual amount of money that the state had available to pay the winner. If the winner had chosen the option of annual payments, the state would have made the first payment and invested the remaining money in some interest-bearing fund from which periodic payments would be paid out. It is similar to the amortized loan situation, where a lump sum of money is transferred from one party to another, and the party receiving the lump sum must make payments to the other party over time. In this particular case, the interest rate that the state is getting on its money is about 2.612% per year. The cash value of the lottery prize is the present value in this situation. The annual payments correspond to the loan payments R in our loan formula except that because one initial payment is made up front, the present value of $3.1 million is given by

$$3,100,000 = R + R\left(\frac{1 - (1.02612)^{-25}}{0.02612}\right)$$

Solving for R, we see that

$$3,100,000 = R\left(1 + \frac{1 - (1.02612)^{-25}}{0.02612}\right)$$

$$3,100,000 \approx R(19.19053303)$$

$$R \approx \frac{3,100,000}{19.19053303}$$

$$\approx \$161,538.$$

When should the winner choose the lump sum option? Ignoring taxes, the winner should take the lump sum if it can be invested at better than 2.612% interest compounded annually. However, taxes will probably be higher on the large lump sum, and that must be taken into consideration. Of course, there are other factors that come into play in choosing a method of payment, for instance the age or any pressing financial needs of the winner.

Some lottery games, for instance New York Lotto and Colorado Powerball, offer the option of a lump-sum payment or a *graduated* payment plan. With the graduated payments, the payments gradually increase in size. In this case, the state can hold some of the money longer so the cash value of the lottery prize is an even smaller fraction of the sum of the scheduled payments.

Not all state lotteries offer a lump-sum option, and some states that do require the players to select that option when the lottery ticket is purchased. Because of this, there are several investment companies that will pay lottery winners lump-sum payments in exchange for their stream of payments.

TECHNOLOGY *tip*

The repetitive nature of constructing amortization tables makes their construction an ideal spreadsheet application.

EXERCISES FOR SECTION 2.5

In Exercises 1–4, find the principal P of a loan with the given payment, time period, and interest rate.

1. $R = \$290.62$, $t = 5$ years, $r = 8\%$ compounded monthly
2. $R = \$1427$, $t = 15$ years, $r = 10.7\%$ compounded quarterly
3. $R = \$90$, $t = 6$ years, $r = 7.4\%$ compounded quarterly
4. $R = \$31.49$, $t = 9$ months, $r = 6\%$ compounded monthly

In Exercises 5–8, find the amount of each payment R for a loan with the given principal, time period, and interest rate.

5. $P = \$105,000$, $t = 20$ years, $r = 7.75\%$ compounded monthly
6. $P = \$4623.14$, $t = 8$ years, $r = 11\%$ compounded quarterly
7. $P = \$7430.65$, $t = 18$ months, $r = 13\%$ compounded quarterly
8. $P = \$87,400$, $t = 10$ years, $r = 8.8\%$ compounded monthly

In Exercises 9–12, find the time t, in years, required to pay off a loan with the given principal, payment, and interest rate.

9. $P = \$16,425$, $R = \$500$, $r = 9\%$ compounded monthly
10. $P = \$60.82$, $R = \$10$, $r = 19.5\%$ compounded monthly
11. $P = \$300,000$, $R = \$12,000$, $r = 7.3\%$ compounded quarterly
12. $P = \$8472$, $R = \$350$, $r = 6\%$ compounded quarterly

13. Suppose you can obtain an 8-year bank loan at an interest rate of 9% compounded quarterly and you have determined that you can afford to make payments of $350 each quarter. How much can you afford to borrow?

14. The balance on a Capital One Visa Card account is $1373.99. The minimum payment due is $25 and the interest rate is 17.9% compounded monthly. Suppose only $25 is paid on the account each month.
    (a) How long would it take to pay off the balance?
    (b) How much total interest is paid by the time the balance is paid off?

15. The Federal Perkins Loan is a low-interest student loan available to students with exceptional financial need. The loan has an interest rate of 5% compounded

monthly with monthly payments and the borrower has up to 10 years to repay. Suppose a student has a $9000 Perkins Loan program and will take 10 years to repay it.
(a) What are the monthly payments?
(b) What is the total amount of payments over the life of the loan?
(c) How much interest is paid over the life of the loan?

16. An Edsel, a classic car from the late 1950s, is offered for $7500. Suppose it is purchased with a 15% down payment and a 3-year loan at 9.5% interest compounded monthly.
(a) What are the monthly payments?
(b) What is the total amount of payments over the life of the loan?
(c) How much interest is paid over the life of the loan?

17. Suppose a student takes out a 10-year $9000 student loan at 7% interest compounded quarterly with quarterly payments. After 8 years of payments, the student decides to pay off the balance of the loan. How much will the student have to pay?

18. A woman took out a 4-year, $15,460.35 car loan at 9% interest compounded monthly with monthly payments. After making payments for 3 years, what is the balance of the loan?

19. To purchase a $23,620 swimming pool, a family makes a 10% down payment and finances the remainder with a loan at 8.3% interest compounded quarterly. If they make loan payments of $2000 per quarter, how long will it take them to pay off the loan?

20. From 2008 to 2009, the average mortgage rate in the United States on a 30-year loan fell from 6.03% to 5.04%. Suppose a family could only afford mortgage payments of $700 per month. How large a loan could they afford if the interest is compounded monthly at a rate of
(a) 6.03%?   (b) 5.04%?

21. The highest average interest rate on 30-year mortgages in the United States from 1990 to 2011 was 10.13% in 1990. In that year the median house price was about $79,100. Suppose a 30-year mortgage was taken out on a $79,100 home in 1990 at an interest rate of 10.13% compounded monthly. Find
(a) the monthly payments,
(b) the balance of the loan in the year 1996,
(c) the balance of the loan in the year 2010.

22. A small company took out a 5-year $20,000 loan at 9.5% interest compounded quarterly with quarterly payments. After 3 years of payments, the company wants to pay off the balance of the loan. How much will it have to pay?

23. A car dealer offered a Toyota Corolla for a total price of $16,738. Assume that a 10% down payment was required. The customer had a choice of taking out a loan for the remaining cost of the car at 0.9% interest compounded monthly or taking a $750 rebate to be used to reduce the size of the loan by $750. If the rebate were taken, then the buyer could only obtain a loan

at 7.95% interest compounded monthly. In either case, the loan required monthly payments for 4 years. For which option are the monthly payments the lowest?

24. Suppose $10,000 is borrowed at 8.7% interest compounded monthly.
    (a) How long will it take to pay off the loan with monthly payments of $200?
    (b) How long will it take to pay off the loan with monthly payments of $100?
    (c) What happens if you use the loan formula to find out how long it will take to pay off the loan with monthly payments of $50? Why does the formula not work in this case?

25. A family takes out a $138,542.91 mortgage at 8.4% interest compounded monthly with monthly payments for 30 years.
    (a) How much would the scheduled payments on the loan be?
    (b) If they decide to pay an extra $200 each month on the house payments, how long will it take to pay off the loan?
    (c) How much will they save in payments over the life of the loan by paying $200 extra each month?

26. Suppose you have a student loan of $5700 at 7% interest compounded quarterly with quarterly payments for 10 years.
    (a) How much would the scheduled payments on the loan be?
    (b) If you pay an extra $75 each quarter, how long will it take to pay off the loan?
    (c) How much will you save in payments over the life of the loan by paying $75 extra each quarter?

27. In January 2012, the average annual rate on a $100,000 mortgage for a 30-year fixed interest rate loan in the United States was 3.92%, compounded monthly. The average rate on a 15-year fixed interest rate loan was 3.20%, compounded monthly. Suppose a mortgage is taken out for $100,000 and the interest is at these average rates.
    (a) Find the monthly payment for the 30-year mortgage.
    (b) Find the monthly payment for the 15-year mortgage.
    (c) How much would a borrower save in total payments over the life of the loan by choosing the 15-year mortgage rather than the 30-year mortgage?

28. A Honda dealership had a 2012 Honda Accord for a total price of $26,583.76. Honda was offering two choices of financing at the time: a 36-month loan at 0.9% compounded monthly and a 48-month loan at 1.9% compounded monthly. Either of the loans would have monthly payments. Suppose the Accord is purchased with a 10% down payment and the remainder is financed with a loan. Find the total amount that would be made in payments and the total interest paid over the life of the loan for each of the two loan choices.

29. An undergraduate takes out an $8000 student loan at 6% interest compounded quarterly with quarterly payments for 10 years. Write an amortization schedule for the first year of the loan.

30. A family borrows $75,350.25 to purchase a home. The loan is at 9.6% interest compounded monthly with monthly payments for 15 years. Write an amortization schedule for the first 6 months of the loan.

31. A man finances the purchase of his living room furniture with a loan of $5429.38 at 8.4% interest compounded monthly with monthly payments for 6 months. Write an amortization schedule for the 6 months of payments.

32. Write out an amortization schedule for a $2500 two-year loan at 7.2% interest compounded quarterly with quarterly payments.

33. Write out an amortization schedule for a $7300 loan at 11.5% interest compounded quarterly with quarterly payments for 9 months.

34. Suppose a $2356.34 computer purchase is financed with a loan at 13.5% interest compounded monthly with monthly payments for 5 months. Write an amortization schedule for the loan.

35. A man borrowed $115,000 to buy a house. The loan he obtained had an interest rate of 9.45% compounded monthly with monthly payments for 30 years. After 10 years of payments, interest rates have dropped and he is considering refinancing by paying off the old loan and taking out a new 20-year loan at 8.5% interest compounded monthly with monthly payments.
    (a) Find the total amount the man would make in payments over the life of the loan for the original loan.
    (b) Find the principal for the new loan.
    (c) Find the total amount the man will make in payments for both loans if he decides to refinance.
    (d) If the bank will charge the man $3000 to refinance, should he refinance? *(Do not consider any interest that might be earned by investing the refinance charge or investing the difference in monthly payments.)*

36. A couple purchased a house with a $237,400.30 loan. The loan had an interest rate of 8.7% compounded monthly with monthly payments for 30 years. After 15 years of payments, they are considering refinancing by paying off the old loan and taking out a new 15-year loan at 7.75% interest compounded monthly with monthly payments.
    (a) Find the total amount that would be made in payments over the life of the loan for the original loan.
    (b) Find the principal for the new loan.
    (c) Find the total amount that would be made in payments for both loans if they decided to refinance.
    (d) If the bank will charge the couple $1500 to refinance, should they refinance? *(Do not consider any interest that might be earned by investing the refinance charge or investing the difference in monthly payments.)*

For Exercises 37 and 38, consider the following: In purchasing a house, the buyers sometimes have the option of paying some extra money up front in order to lower the interest rate they will pay for their mortgage. The money paid up front is called *points*, and each point costs the buyer 1% of the total amount of the mortgage. (*In all parts of these questions, do not consider any interest that might be earned by investing the point or investing any difference in monthly payments.*)

37. In January 2012, Vanguard Mortgage offered 30-year home mortgages at a rate of 3.99% with no points or 3.625% with one point. The interest in both cases was compounded monthly, and the loan was to be paid back with monthly payments for 30 years. Suppose a man borrows $195,000 from Vanguard Mortgage to purchase a home when these rates are in effect.

(a) If the loan is to be held for the entire 30 years, how much will paying the point save in payments over the life of the loan?

(b) If the loan is to be held for the entire 30 years, should the man pay the point?

(c) Suppose the man plans to sell the house after only 2 years. How much will paying the point save in payments over the 2 years?

(d) What is the balance on the loan after 2 years if the point is paid?

(e) What is the balance on the loan after 2 years if the point is not paid?

(f) If the man plans to sell the house in 2 years, should he pay the point?

38. In January 2012, Quicken Loans offered 15-year home mortgages at 3.5% with no points and at 3.25% with one point. The interest in both cases was compounded monthly, and the loan was to be paid back with monthly payments. Suppose a woman plans to take out an $203,500 mortgage to purchase a condominium when these rates are in effect.

(a) If the loan is to be held for the entire 15 years, how much will paying the point save in payments over the life of the loan?

(b) If she plans to hold the loan for the entire 15 years, should she pay the point?

(c) If she plans to sell the condominium in 5 years, how much will paying for the point save her in payments over the 5 years?

(d) What is the balance on the loan after 5 years if she pays the point?

(e) What is the balance on the loan after 5 years if she does not pay the point?

(f) If she plans to sell the condominium in 5 years, should she pay the point?

39. In Exercise 37, we ignored any interest that might have been earned by investing the point or any difference in monthly payments in a savings account. Suppose that in the situation in Exercise 37, either the borrower does not pay the point and invests the money he would have paid for the point in an account earning 3% interest compounded monthly, or he does pay the point and each month he invests the reduction in the monthly loan payment into an account earning 3% interest compounded monthly.

(a) If the borrower plans to hold the loan for the entire 30 years, should he pay the point?

(b) If the borrower plans to hold the loan for only 2 years, should he pay the point?

40. In Exercise 38, we ignored any interest that might have been earned by investing the point or any difference in monthly payments into a savings account. Suppose that if the woman does not pay the point, she invests the money she would have paid for the point in an account earning 5% interest compounded monthly, and if she does pay the point, then each month she invests the reduction in the monthly loan payment into an account earning 5% interest compounded monthly.

(a) If the woman holds the loan for the entire 15 years, should she pay the point?

(b) If the woman holds the loan for only 5 years, should she pay the point?

For Exercises 41–46, consider the following: There are some situations in which a lump sum is invested into an account earning interest and regular payments are made from the account until it is depleted. This kind of account is often called an annuity, although the word annuity has some other alternative meanings. If the lump sum invested is P, interest is compounded n times per year at an annual rate r, and payments of size R are paid out of the account at the end of every period until the account runs out of money at the end of t years, then the loan formula applies to this situation. In this case, we view the bank or holder of the account as the borrower and the person receiving the payments as the lender.

41. Suppose you want to set up an account from which $500 payments can be withdrawn at the end of each month for 15 years. If the account will earn 7.25% compounded monthly, what lump sum would need to be invested at the start?

42. A student will be attending college for the next 4 years, and her parents want to set up an account from which $3000 payments for college expenses can be withdrawn at the end of each quarter of the next 4 years. If the account will earn 6.8% interest compounded quarterly, how much should be invested?

43. Suppose a woman's retirement account currently has $257,548.72 in it and it is earning 6.72% interest compounded monthly. If she plans to withdraw $4000 from the account at the end of each month, how long will it take for the funds in the account to be depleted?

44. A grandmother wants to set up a trust fund account from which payments to her grandson can be withdrawn at the end of each month for the next 20 years. If $80,000 is invested in the account and it earns 7% interest compounded monthly, what size will the payments be?

45. In a "Hole-In-One Shootout" benefiting the Lena Pope Home, the grand prize was $1,000,000 to be paid in equal monthly installments over 30 years. Assuming the payments will be paid at the end of each month out of an account earning 6.5% interest compounded monthly, what lump sum would need to be invested at the start?

46. Emmitt Smith, the NFL's all time rushing leader, signed a contract with the Dallas Cowboys in 1996 paying a signing bonus of $1.5 million for each of the succeeding 7 years.

(a) Assuming the Cowboys could have invested the money into an account earning 7% interest compounded annually, what lump sum would they need to have invested into an account in 1996 so that Emmitt Smith's $1.5 million payments could have been withdrawn at the end of each year for the following 7 years?

(b) Earlier in the contract negotiations the Cowboys had announced a signing bonus of $15 million spread out over "many years." Assuming the lump sum they planned to invest was the one you found in part (a) and the payments would be made out of an account earning 7% interest compounded annually, with payments made to Emmitt Smith at the end of each year, estimate the number of these "many years" if the total of the payments were to come to $15 million. (You cannot solve for the time exactly here; instead, you will have to estimate it by using a calculator or computer that will perform such estimations or by making guesses that will close in on the answer.)

For Exercises 47 and 48, consider the following: As described in Branching Out 2.7, when a person wins a state lottery and the winnings are paid out with equal annual payments over many years (typically, between 20 and 30 years), the state does not have the entire jackpot on hand but instead invests a smaller amount of money into a fund earning interest, from which the payments to the winner will be paid. If the lump sum invested is P, interest is compounded n times per year at an annual rate r, and payments of size R are paid out of the account with one payment immediately and then at the end of every period until the account runs out of money at the end of t years, then the loan formula applies to this situation except that, because one initial payment is made up front, the present value P is given by

$$P = R + R\left(\frac{1 - \left(1 + \frac{r}{n}\right)^{-nt}}{\frac{r}{n}}\right)$$

47. In the Tri-State Megabucks Plus lottery run by the three states Maine, New Hampshire, and Vermont, winners may collect the jackpot in one of two ways: either in 30 equal annual payments or with a one-time lump-sum cash payment up front. If the payment option is selected, the first of the 30 payments is paid immediately and the remaining payments are paid at the end of each year for the next 29 years. Suppose the Tri-State Megabucks Plus lottery now has $3 million available for a lottery prize. Suppose the states can invest this amount into an account earning 6.82% compounded annually, and the money from this account can be used to make the 30 payments if the winner chooses the payment option. For this arrangement, find

(a) the amount of each annual payment,

(b) the total amount paid out in payments (this is the amount that the state will announce as the jackpot amount).

48. The Florida Lotto gives the winners the option of choosing a lump-sum cash payment or collecting the entire jackpot in 30 equal annual payments with the first payment made right away and the remaining 29 payments paid at the

end of each year for the next 29 years. On December 24, 2011, the jackpot was $50 million and the lump-sum cash option was $34,671,916.67. Suppose you could invest the lump sum in an account earning 7% interest compounded annually and then make 30 equal payments to yourself out of the account in the same way as the state's payment plan so that the account would run out of funds with the last payment.

(a) Find the amount of each payment.

(b) If you play, should you choose the lump-sum option or the payment option?

49. An amortized loan may have interest compounded more often than payments are made. For instance, quarterly payments might be made on a loan for which interest is compounded monthly. The amortized loan formula derived in the text does not directly apply to this case, but it can be adapted.

(a) Find an amortized loan formula for a loan with principal P, an annual interest rate r, compounded n times a year, for which payments of size R are made at the end of every k periods for t years. (Assume nt is exactly divisible by k.)

(b) Use the formula from part (a) to find the payment on a $5000 loan at 8% interest compounded monthly with payments made at the end of each quarter for 4 years.

50. Although it is not possible to algebraically solve exactly for the interest rate r in the amortized loan formula, r can be estimated by use of a calculator or computer that will perform such estimates or by making a series of guesses for r. When using the guessing method, you should determine whether your first guess was too high or too low and adjust accordingly, and continue to adjust your guesses until the value for r you have found gives a reasonably close answer. Suppose that the payments on a $20,000 loan with interest compounded monthly and monthly payments for 5 years are $526. Use some method to estimate the annual interest rate r to two decimal places.

WRITING EXERCISES

1. Would you expect the variation in interest rates among different lenders to be greater now or in the past? Give a detailed argument supporting your position.

2. Collateral, credit history of the loan applicant, the length and size of a loan, and the purpose of a loan all affect the interest rates banks charge. Discuss how these factors influence interest rates. Can you think of other factors that influence rates as well?

3. Unlike the fixed-rate mortgages we have been considering, interest rates on an adjustable-rate mortgage (ARM) may change over the life of the loan. Typically, they begin at a rate one or two percentage points below that of a fixed-rate mortgage. Each year the rate is adjusted (up or down) to be some set level above the U.S. government's current prime interest rate. The consumer is usually somewhat sheltered from huge increases in the rate by caps on the increase in

any one year and by a ceiling on the highest rate over the life of the loan. Discuss different circumstances in which the initial lower interest rate of an ARM is worth the additional risk of higher rates in the future.

4. Suppose you are a financial planner and are seeking new clients. Write a letter that you will send to families with small children in which you are trying to impress on them the importance of early and careful financial planning in saving for college. Of course, you want to persuade them to come to you for help.

PROJECTS

1. Study the effect of changing interest rates on the affordability of homes. Research the historical relation of interest rates to number of home purchases and housing starts.

2. Compare the traditional IRA and the Roth IRA. Under what circumstances should an investor prefer one over the other? You should take into account the tax bracket of the investor before and after retirement and how long the money will be invested.

3. Choose a state with a lottery in which players can choose to have their winnings paid in a lump sum or in a series of payments. Taking into consideration not only the payments but also state and federal income taxes, analyze the question of which option a player should choose.

4. Compare the financial aspects of leasing a new car versus buying the same car.

5. Create a spreadsheet or computer program that will compute the amortization table for a loan.

KEY TERMS

amortization schedule, 133
amortized loan, 126
annual percentage yield (APY), 103
compound interest, 99
compound interest formula, 101
future value, 94

inflation, 109
loan balance, 131
loan formula, 127
period, 100
periodic interest rate, 100
present value, 94, 126

principal, 92, 126
simple interest, 92
simple interest formula, 93
simple interest future value formula, 94
systematic savings plan, 115
systematic savings plan formula, 118

CHAPTER 2 / REVIEW TEST

1. The average rate of interest on a 6-month certificate of deposit in the United States in August 1981 was a very high 17.98% and in February 2012 it was only 0.52%. How much more interest would be earned on a $10,000 6-month certificate of deposit at 17.98% compounded monthly than one at 0.52% compounded monthly?

2. A loan shark offers to lend $500 on the condition that he is paid back $800 at the end of 9 months. What simple interest rate is the loan shark charging?

3. Suppose a 22-year-old man starts to make $500 deposits at the end of each quarter into an account earning 7% compounded quarterly.

    (a) How much will be in the account when the man is 68 years old?

    (b) How much of the future value will be from deposits?

    (c) How much of the future value will be from interest?

4. Suppose a 7-year loan at 8.2% compounded monthly with monthly payments is used to purchase a 2012 Rolls Royce for $223,147.

   (a) What are the monthly payments?

   (b) What are the total payments over the life of the loan?

   (c) How much interest is paid over the life of the loan?

5. Based on the Consumer Price Index, the rate of inflation in the United States between 1940 and 2010 was about 3.9% per year. What salary in 1940 would be equivalent to a $50,000 salary in 2010?

6. If $3000 is invested at 7% interest compounded quarterly, how long will it take until the investment is worth $10,000?

7. From January 2008 to January 2012, the average mortgage rate in the United States on a 30-year loan fell from 6.25% to 4.24%. Suppose a family could only afford mortgage payments of $1200 per month. How large a loan could they afford if the interest is compounded monthly at a rate of

   (a) 6.25%?

   (b) 4.24%?

8. A couple wants to save $15,000 to be used toward the down payment on a home. If they can invest money in an account earning 7.3% interest compounded quarterly, how much do they have to deposit at the end of each quarter in order to save $15,000 in 4 years?

9. Queen Elizabeth I was the first English monarch to enforce the maximum legal interest rate of 10% that had been established 15 years earlier by King Edward VI. In 1561, she borrowed 30,000 pounds in London at 10% interest. How much simple interest would she have paid by 1570?

10. A couple borrows $943.61 to pay for a refrigerator. The loan is at 7.8% interest compounded monthly with monthly payments for 6 months. Write an amortization schedule for the loan.

11. A man takes out a $18,493.05 car loan at 9.9% interest compounded monthly with monthly payments for 5 years. What is the balance of the loan after

    (a) 1 year?

(b)   4 years?

12. In March 2012, Chase Bank offered a 12-month certificate of deposit at an interest rate of 0.40% compounded monthly. If $30,000 were invested in this CD, how much would it be worth after 1 year?

13. Suppose a small company takes out a loan of $14,000 at 8.4% interest compounded monthly with monthly payments for 10 years.

    (a)   How much would the scheduled payments on the loan be?

    (b)   If the company decides to pay an extra $50 each month on the loan payments, how long will it take to pay off the loan?

(c)   How much will the company save in payments over the life of the loan by paying $50 extra each month?

14. Suppose you invest $500 and would like the investment to grow to $10,000 in 16 years. What interest rate, compounded daily, would you have to earn in order for this to happen?

15. On March 25, 2012, Wells Fargo paid 1.15% interest compounded monthly on a 4-year certificate of deposit with a minimum deposit of $5000. Find the annual percentage yield (APY).

16. To save $160 to use toward the purchase of a video game system, a boy deposits $10 at each month's end into a savings account earning 4% interest compounded monthly. How long will it take before the account is worth $160?

17. If $3800 is borrowed at 5% interest compounded quarterly and payments of $200 are made each quarter, how long will it take to pay off the loan?

18. In April 2011, Arundel Federal Savings Bank offered 4-year new car loans at 6.4% interest if a 10% down payment was made or at 5.4% interest if a 20% down payment was made. In both cases the interest is compounded monthly and payments are made each month for 4 years. Suppose a $28,000 car is purchased. Find the total amount paid (including the down payment) for the car on each of these two financing options.

19. A woman deposits $300 at each quarter's end into a savings plan account earning 4.5% compounded quarterly. After 3 years, the interest rate drops to 4% compounded quarterly and she stops making deposits. If she leaves the money in the account, how much will the account be worth 8 years after she opened it?

20. The Virginia Prepaid Education Plan allows families to prepay college tuition for ninth graders and younger children. In 2012, a prepaid in-state four-year university tuition contract for an infant, who would be expected to enter college in 18 years, could be purchased for a lump sum of $56,600. For those who chose the option of making monthly payments from the time the child was an infant until he or she entered college 18 years later, the cost was $483 per month. The

tution and fees for 4 years at the University of Virginia in 2012 were $46,336, and tution from 1970 to 2012 had been increasing at a rate of about 8% per year.

(a) Assuming that tuition at the University of Virginia continues to increase at a rate of 8% per year, how much will tuition be 18 years after 2012, in 2030?

(b) Assuming tuition will cost the amount you found in part (a) in 2030, what is the annual interest rate, compounded monthly, that the purchasers of the lump-sum contract in 2012 earn on their $56,600 investment?

(c) Find the future value of the $483 monthly payments for 18 years at the interest rate you found in part (a), compounded monthly. Assume the payments are made at the end of each month.

(d) Which of the payment options is a better deal for the purchasers?

SUGGESTED READINGS

Groz, Marc M. *Forbes Guide to the Markets: Becoming a Savvy Invester.* New York: John Wiley, 2009. Comprehensive guide to investment strategies, including investing in stocks, mutual funds, and bonds, and a discussion of risks versus returns.

Nissenbaum, Martin, Barbara J. Raasch, and Charles L. Ratner. *Ernst & Young's Personal Financial Planning Guide.* New York: John Wiley, 2004. Very detailed personal finance guide, including a thorough coverage of tax issues. One of many such books.

Wade, William W. *From Barter to Banking: The Story of Money.* New York: Crowell-Collier Press, 1967. Early history, checks, the Federal Reserve, inflation, and the world monetary network.

Williams, Jonathan, et al., eds. *Money: A History.* New York: St. Martin's Press, 1997. Colorful survey of the history of coins and currency.

Creating communications networks or social networks, planning transportation routes, and designing manufacturing processes often involve thousands or even millions of pieces of information. The mathematical translation of the questions of optimal solutions to such problems brings us to graph theory, the focus of this chapter. Graph theory was launched by Leonhard Euler in 1736, when he analyzed the puzzle of the Königsberg bridges. Today, graph theory is routinely used in business and manufacturing. Computers, along with mathematical advances, have made it possible to solve some graph theory problems of enormous magnitude. The scope and complexity of problems in communications, transportation, and manufacturing would overwhelm the fastest foreseeable computers unless they are paired with a mathematical approach that reduces the scale of such problems. The rewards of applying this kind of combination of mathematics and computer science have included savings of billions of dollars over the past 30 years.

Suppose we want to construct the least-expensive network linking a group of computers. Perhaps we need to construct the shortest route along city streets for mail delivery or garbage pickup. Suppose a salesman wants to visit a group of cities in the shortest possible round-trip. Each of these problems may be formulated in terms of graph theory. In this chapter, we examine these three problems. Collectively, they show some of the major ideas of applied graph theory. We will demonstrate a fast and simple solution to the computer network problem. Although there also exists a fast and effective method for solving the problem of mail delivery or garbage pickup, its complexity allows us only a glimpse of this solution. Finally, finding a fast method of determining the best route for the traveling salesman's trip is one of the most famous unsolved problems in mathematics. It is quite possible that no fast method exists. Currently we must be satisfied with finding solutions that are nearly optimal.

PATHS AND NETWORKS

CHAPTER 3

3.1   EULERIAN PATHS AND CIRCUITS ON GRAPHS

In the eighteenth century, the Prussian town of Königsberg (now called Kaliningrad) was divided by the Pregel River, as shown in Figure 3.1. The town included the two islands in the middle of the river and land on each bank of the river. The islands and land were connected to one another by seven bridges.

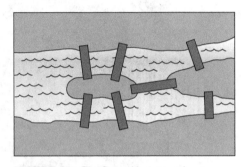

FIGURE 3.1 The Prussian town of Königsberg.

Some of the townsfolk amused themselves by trying to find a path through the city that would cross each bridge exactly once. The path could start anywhere and need not end where it started. Try to trace out such a path yourself and you will see that it does not seem possible. This problem came to be known as the Königsberg bridge problem. The Swiss mathematician Leonhard Euler, then at St. Petersburg, considered the problem and published a paper in 1736 that not only proved that no such path existed but also laid the foundations for a new area of mathematics called graph theory. Euler's great contribution was to see how the complex picture of the map could be replaced by a simpler picture that captured the essence of the problem.

Present day Kaliningrad, formerly known as Königsberg.

To solve the Königsberg bridge problem, Euler replaced each piece of land by a point and each bridge by a curve joining the points. In Figure 3.2, the four different pieces of land are labeled with points *A*, *B*, *C*, and *D* and curves are drawn between the points along each of the bridges.

This map of land and bridges can be viewed more simply by erasing the underlying picture and smoothing out the curves. We are left with the simple figure in Figure 3.3.

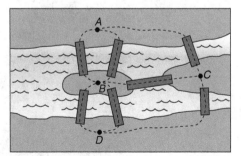

FIGURE 3.2

Figure 3.3 is an example of a graph. In general, a **graph** consists of points called vertices (a single point is called a **vertex**) and lines or curves starting and ending at vertices called **edges.** The mathematical subject of **graph theory** is the study of graphs and their applications. Before looking at Euler's solution to the Königsberg bridge problem, we introduce a few general graph theory definitions.

An edge of a graph may connect two different vertices or it may start and end at the same vertex. If an edge starts and ends at the same vertex, it is called a **loop**. In Figure 3.4, the edge beginning and ending at vertex A is a loop.

A vertex with an odd number of edges attached to it (a loop at a vertex is counted twice) is called an **odd vertex**. A vertex with an even number of attached edges is called an **even vertex**. For instance, in the graph in Figure 3.5, the vertices B, C, and D are even: vertices B and D have four

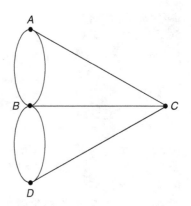

FIGURE 3.3 Königsberg bridge problem graph.

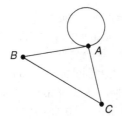

FIGURE 3.4 A graph with a loop at vertex A.

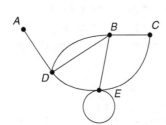

FIGURE 3.5

edges attached to them and vertex C has two edges attached to it. The vertices A and E are odd because A has one edge attached to it and E has five edges attached to it, where the loop at E has been counted twice.

When drawing graphs it is sometimes convenient or necessary to have edges that cross over one another, as they do in the graph shown in Figure 3.6. Note here that you can tell that the point at which the edge between A and D and the edge between B and C cross is not a vertex because we will always indicate vertices by dots. We will not consider these two edges to be intersecting, but instead think of one of the edges as passing above the other.

A **path** on a graph is a route along the edges that starts at a vertex and ends at a vertex. In Figure 3.7, a path from vertex A to vertex B is marked in blue. The direction of

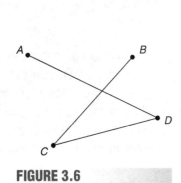

FIGURE 3.6

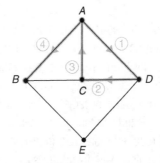

FIGURE 3.7 A path from vertex A to vertex B.

branching OUT

GRAPH THEORY AND THE SOLUTION OF THE FOUR COLOR PROBLEM

How many different colors would be needed to color any map in such a way that no two regions, for instance countries or states, with a common border are colored with the same color? (Countries that only touch at single points are not considered to have a common border.) To color the map of South America, which includes Brazil, Bolivia, Paraguay, and Argentina, we see that we would need four colors because each of these four countries has a common border with each of the other three.

The question of whether or not four colors would suffice for coloring any map was first formally asked in 1852 by Francis Guthrie, a graduate student at University College in London, and the problem eventually became known as the Four Color Problem. Even though no map could be found that required more than four colors, a proof that four colors would work for every map

remained elusive in spite of the efforts of many mathematicians. The fact that four colors appeared to be enough did not ensure that they were, in fact, enough. Perhaps there was some strange map that no one had thought of that would require more than four colors.

The Four Color Problem can be viewed as a problem in graph theory by translating the map into a graph. This is done by representing each country by a vertex, and joining two vertices with an edge if and only if the corresponding countries have a common border. For instance, we see here how the map of five states is translated to a graph.

Image © Jennifer Pavelski, 2012. Shutterstock, Inc.

The problem of coloring the map can be seen as the problem of coloring the vertices of the corresponding graph in such a way that no two vertices connected by an edge have the same color. The vertices of the graph corresponding to the map of the five states are colored in this way. Using this

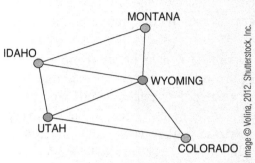

graph-theoretical approach, American mathematicians Kenneth Appel and Wolfgang Haken proved in 1976 that four colors would suffice to color any map. Their proof enlisted the help of a computer program that checked approximately 1500 special cases, requiring about 1200 hours of computer time. To date, no one has found a proof that does not make use of computer calculations.

the path is indicated by the arrows and the edges have been numbered in the order in which they are traveled. A path that begins and ends on the same vertex is called a **circuit**. A graph is **connected** if for any two of its vertices there is at least one path connecting them. Thus, a graph is connected if it consists of one piece. If a graph is not connected, it is **disconnected.** The graph in Figure 3.8(a) is connected, whereas the graph in Figure 3.8(b) is disconnected.

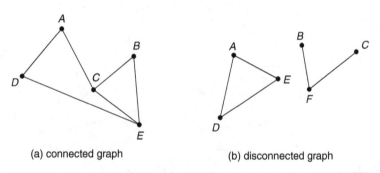

(a) connected graph   (b) disconnected graph

FIGURE 3.8 (a) connected graph (b) disconnected graph

In the Königsberg bridge problem, we are trying to find a path that traverses each edge of the graph exactly once. We have the following important definitions:

A path that includes each edge of a graph exactly once is called an **Eulerian path**, and an Eulerian path that is also a circuit is called an **Eulerian circuit.**

The path from D to C in Figure 3.9 is an Eulerian path because it includes each edge of the graph exactly once. See if you can find some of the other Eulerian paths in the graph.

The path shown in Figure 3.10 is an Eulerian circuit because it includes each edge of the graph exactly once and it begins and ends at the same vertex, A. You should try to find some of the other Eulerian circuits.

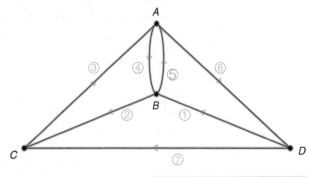

FIGURE 3.9

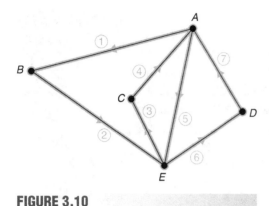

FIGURE 3.10

We now see that the Königsberg bridge problem is the problem of finding an Eulerian path on the graph in Figure 3.3.

Eulerian paths and circuits have many practical applications. In garbage collection, the workers must complete a route that covers every street in an area. For maximum efficiency, it is desirable to cover each street exactly once. In mail delivery or reading parking meters, a worker may need to cover both sides of some streets, but these routes can also be modeled by graphs for which we want to find Eulerian paths or circuits. We look at some applications of this type later in this section and in the exercises.

To understand how Euler analyzed the Königsberg bridge problem, we consider the graph in Figure 3.11. We will see that this graph has an Eulerian path but not an Eulerian circuit. The path in Figure 3.12 shows an Eulerian path. So far, so good.

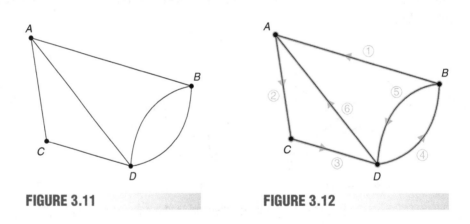

FIGURE 3.11   **FIGURE 3.12**

What happens if you try to trace out an Eulerian circuit on the graph? After many tries, you might guess that there is none. Let's look at the Eulerian path we discovered for clues to what is happening. Edges meeting at a vertex, except for the vertices at the beginning and end of the path, are covered two at a time: one on the way into the vertex and one on the way out. This is a general property holding for every Eulerian path or circuit. Therefore, the vertices in a graph for which an Eulerian path or circuit exists must all be even, with the possible exception of the initial and final vertices. In an Eulerian circuit, the path begins and ends at the same vertex, so that this vertex will also be even. In an Eulerian path with different initial and final vertices, both of these vertices will be odd. Therefore, there cannot be an Eulerian circuit on a graph unless all vertices are even and there cannot be an Eulerian path on a graph unless all the vertices are even or only the initial and final vertices are odd. Euler discovered this and stated that these conditions are enough to guarantee the existence of an Eulerian circuit or path, although it took over 100 years for a formal proof of the existence to be published by mathematician Carl Hierholzer. This result came to be known as **Euler's Theorem,** which we will now formally state.

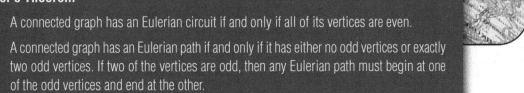

Euler's Theorem

(a) A connected graph has an Eulerian circuit if and only if all of its vertices are even.

(b) A connected graph has an Eulerian path if and only if it has either no odd vertices or exactly two odd vertices. If two of the vertices are odd, then any Eulerian path must begin at one of the odd vertices and end at the other.

Note that if a graph has an Eulerian circuit, then it automatically has an Eulerian path, because the Eulerian circuit is just a special kind of Eulerian path.

We now return to the problem we began with—namely, whether or not it is possible to find a path through the city of Königsberg that crosses each bridge exactly once. As we have seen, this is equivalent to the question of whether or not there exists an Eulerian path on the graph in Figure 3.13.

Because each of the four vertices in the Königsberg bridge graph in Figure 3.13 is odd, Euler's Theorem tells us that there is no Eulerian path on this graph, and therefore it is impossible to walk through the city of Königsberg along a path that crosses each bridge exactly once.

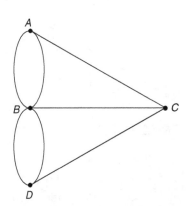

FIGURE 3.13 Königsberg bridge problem graph.

In Example 1, we look for Eulerian paths and circuits. If we are trying to find an Eulerian circuit, then we can start on any vertex. However, if we are trying to find an Eulerian path on a graph with two odd vertices, we must start the path at one of the odd vertices.

EXAMPLE 1 *Eulerian Paths and Circuits*

For each graph in Figures 3.14, 3.15, and 3.16, determine whether the graph has an Eulerian circuit, and if it does, give one. If the graph does not have an Eulerian circuit, determine whether it has an Eulerian path, and if it does, give one.

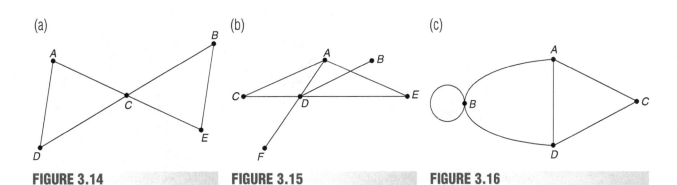

FIGURE 3.14　　　　**FIGURE 3.15**　　　　**FIGURE 3.16**

(a) The vertices of this graph are all even, so we know it has an Eulerian circuit. One circuit is shown in Figure 3.17.

(b) The four vertices *A*, *B*, *D*, and *F* are all odd, so by Euler's Theorem we know that the graph has neither an Eulerian circuit nor an Eulerian path.

(c) Vertices *B* and *C* are even, whereas vertices *A* and *D* are odd. The graph does not have an Eulerian circuit, but it does have an Eulerian path starting and ending at the odd vertices. One such Eulerian path is shown in Figure 3.18.

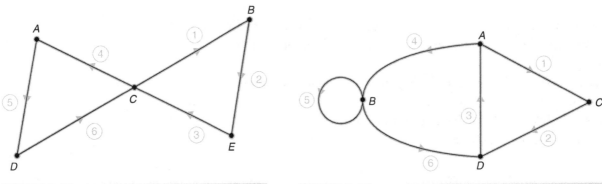

FIGURE 3.17   **FIGURE 3.18**

Note that if a graph has one Eulerian path or circuit, it may have several different ones. For small graphs, using trial and error to find Eulerian paths or circuits works fine. However, with a larger graph a somewhat more systematic approach might be helpful. One such method simply requires that, at each step in tracing out a path or circuit, we should never cover an edge that will cause the uncovered parts of the graph to become disconnected. This method is illustrated in the next example.

EXAMPLE 2 *Finding an Eulerian Path*

Find an Eulerian path in the graph in Figure 3.19.

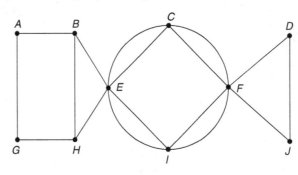

FIGURE 3.19

Because the graph has exactly two odd vertices, *B* and *H*, we know that it has an Eulerian path but not an Eulerian circuit. We start at vertex *B* (the other odd vertex,

H, would be okay too). Beginning with the path marked in Figure 3.20, we have chosen edges that have kept the uncovered part connected up to this stage.

At this point, we need to make a careful choice about what to do next. If we now choose to move back to vertex *B*, the uncovered parts will be disconnected. Therefore, we should choose any of the other edges going out from vertex *E*. We will choose to go to vertex *I* next. Continuing in this way, and being careful not to choose edges that will cause the uncovered part to become disconnected, one possible path is shown in Figure 3.21.

You should trace this out yourself to see that it works. You might also try to find some of the other Eulerian paths in the graph. ■

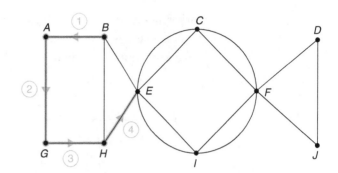

FIGURE 3.20

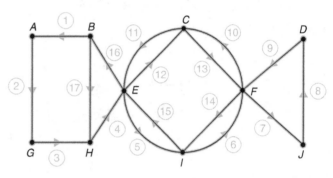

FIGURE 3.21

We will now apply graph theory to a practical problem. The first step is to find a graph that models the problem we are trying to solve. This idea is illustrated in the next example.

EXAMPLE 3 *Postal Worker Route*

Suppose a postal worker must deliver mail to the four-block neighborhood shown in Figure 3.22. She plans to park her truck at one of the street intersections and deliver mail to each of the houses. The streets on the outside of the neighborhood have houses on only one side, and the interior streets have houses on both sides. She must walk down each of the interior streets twice (once for each side) and each of the outside streets once. Draw a graph corresponding to the walking route she needs to cover. Determine whether or not the graph has an Eulerian circuit, and if so, find one.

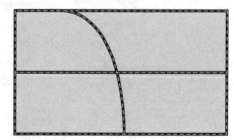

FIGURE 3.22 Street layout of the neighborhood.

SOLUTION:

In the graph corresponding to this problem, each street intersection in the neighborhood is a vertex. The streets with houses on only one side must be covered only once, so they are represented by one edge. The streets with houses on both sides must be covered twice, so they are represented by two edges. We ignore the lengths of the streets and how they are curved. The graph corresponding to the route the postal worker needs to cover is shown in Figure 3.23. Because each of the vertices in this graph is even, we know that it has an Eulerian circuit. There are many possibilities; find a few yourself. One possible Eulerian circuit is shown in Figure 3.24.

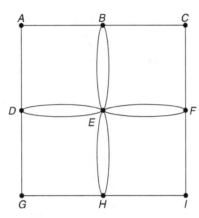

FIGURE 3.23 Graph model of the walking route.

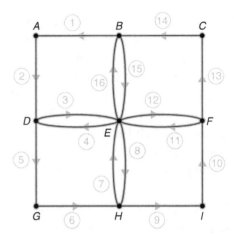

FIGURE 3.24

Eulerization

Suppose a newspaper delivery person must deliver papers in a neighborhood with streets laid out as in the graph in Figure 3.25. Because he delivers papers to houses on both sides of the street at once, he would like to travel down each street exactly once.

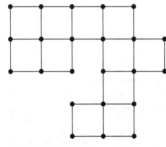

FIGURE 3.25 Street layout of the neighborhood.

He wants to begin and end his route at his home, which lies on a corner in the neighborhood. The question of whether or not such a route exists is equivalent to the question of whether or not the graph has an Eulerian circuit. Because the graph has several odd vertices, we know by Euler's Theorem that such a route is not possible.

To cover all of the streets of the neighborhood, the newspaper delivery person will have to go down some streets more than once. But how can he do this efficiently?

Our newspaper route example leads to the more general question of how to cover all of the edges in a connected graph having some odd vertices in such a way that the number of edges that are reused is kept small. We can visualize the reused edges by altering the original graph, adding new edges corresponding to the edges we will reuse. These new edges are colored green to signify that they were not part of the original graph and will always represent duplicating an existing edge. To get a new graph that has an Eulerian circuit, we want to add new edges in such a way that the altered graph has only even vertices. This process of adding edges in order to guarantee an Eulerian circuit in the new graph is called **eulerization.** *It is essential to note that in eulerization, we are only allowed to add new edges between vertices that have an edge between them in the original graph.*

To see how eulerization works, consider the simple graph in Figure 3.26(a). Because this graph has two odd vertices, it cannot have an Eulerian circuit. However, if an edge is added between the odd vertices A and C, then the resulting graph, shown in Figure 3.26(b), has all even vertices and therefore has an Eulerian circuit.

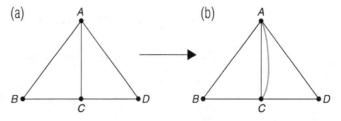

FIGURE 3.26 Adding an edge.

Note that traveling along the added edge on the altered graph actually represents reusing the edge in the original graph. Another way to eulerize the graph would be to add edges between vertices A and B and between vertices B and C, but this would require adding two new edges.

Ideally, in eulerizing a graph, we want to minimize the total length of new edges, for this would minimize the length of the complete circuit. For instance, in the case of the newspaper route, we want to minimize the length of the entire route. The problem of finding such an eulerization is called the *Chinese postman problem*, and it was first studied by the Chinese mathematician Meigu Guan. However, for simplicity, we will not worry about the lengths of the edges and instead will try to keep the total number of new edges to a minimum. This problem is sometimes called the *simplified Chinese postman problem*. Note, however, that given any graph, if we break down the edges into smaller edges, so that all of the edges of the graph have approximately equal length, then the problem of minimizing the total number of new edges is essentially equivalent to the problem of minimizing the total length of the new edges.

There is an algorithm for eulerizing a graph while keeping the number of new edges to a minimum, but it is complicated in the general case, so we will simply use the following guidelines in finding eulerizations:

Guidelines for Eulerizing a Graph

(1)   Circle all of the odd vertices.

(2)   Pair each odd vertex off with another odd vertex that is close to it in the graph.

(3)   For each pair of odd vertices, find the path with the fewest edges connecting them in the original graph, and then duplicate the edges along this path.

A bit of exploration will convince you that there are always an even number of odd vertices, so it is possible to pair them off as required in the first step. These guidelines for eulerizing may not give the best eulerization because it is not always easy to see the best way to pair off the odd vertices. However, the guidelines work reasonably well for finding a good eulerization.

EXAMPLE 4 *Eulerization*

Find an eulerization of the graph in Figure 3.27.

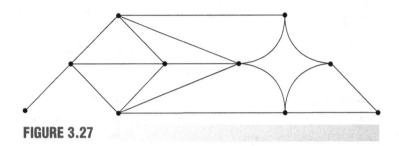

FIGURE 3.27

We first circle all of the odd vertices as in Figure 3.28.

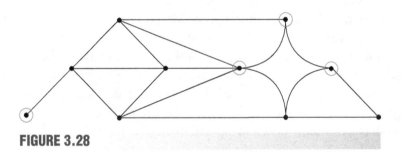

FIGURE 3.28

Pairing the two rightmost odd vertices, pairing the two odd vertices to the left of them, and duplicating edges along a shortest path between each of these pairs, we get the eulerization of the graph shown in Figure 3.29.

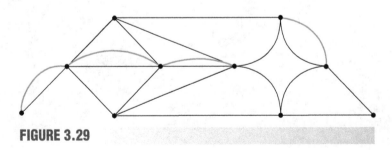

FIGURE 3.29

There are two other paths with just three edges between the two odd vertices on the left (find them), and therefore two other ways to eulerize the graph by duplicating only four edges. To see that this is the minimal number of edges we need to add, first note that the odd vertex farthest to the left is at least three edges away from any of the other three odd vertices, and connecting the remaining two odd vertices will require adding at least one more edge. ■

Let's now return to our newspaper route question and use eulerization to find a reasonable route.

EXAMPLE 5 *Newspaper Route*

Find an eulerization of the graph corresponding to the neighborhood shown in Figure 3.25 that will allow an Eulerian circuit that would be a reasonably efficient newspaper route.

SOLUTION:

First, we circle the odd vertices of the graph corresponding to the neighborhood as in Figure 3.30. One possible eulerization is shown in Figure 3.31.

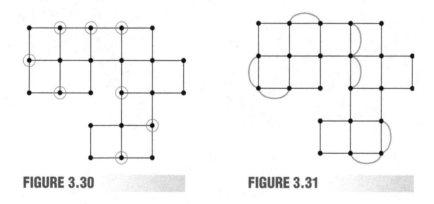

FIGURE 3.30　　　　**FIGURE 3.31**

The eulerized graph in Figure 3.31 has an Eulerian circuit that corresponds to an efficient newspaper route. ■

In applications involving Eulerian paths or circuits, there may be additional complications we have not considered. For instance, it may very well be that the route that a truck must cover includes some one-way streets. Another complication that arises with services, such as street sweeping or snow removal, is that the vehicles must travel down only one side of the street at a time in the direction of traffic. These issues and others must be dealt with when they arise in practice.

EXERCISES FOR SECTION 3.1

In Exercises 1–4, circle all of the odd vertices of the graph.

1.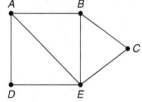

2.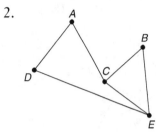

3.

4.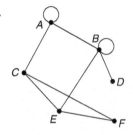

In Exercises 5–14, determine whether the graph has an Eulerian circuit, and if it does, give one. If it does not, explain why not. If the graph does not have an Eulerian circuit, determine whether it has an Eulerian path, and if it does, give one. If it does not, explain why not.

5.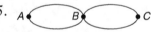

6.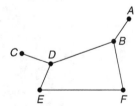

7.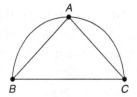

8.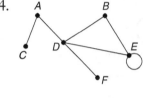

9.

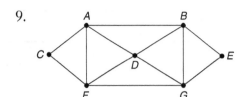

10.

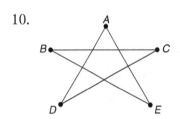

11.

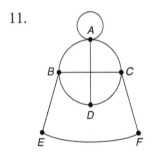

12.

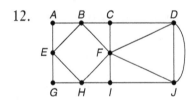

13.

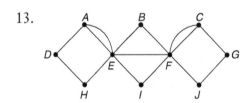

14.

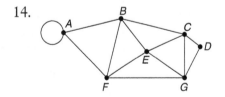

15. Today the city of Königsberg is called Kaliningrad, and in Kaliningrad there are two more bridges in addition to those in the original Königsberg bridge problem. A picture of the current situation is shown here. With these additional bridges, we can again ask if it is possible to start somewhere in the town and walk about the town crossing each bridge exactly once. The walk does not have to start and end at the same place. Draw the corresponding

graph with the pieces of land as vertices and the bridges as edges, and answer this question.

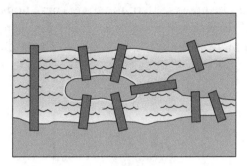

16. In his article on the Königsberg bridge problem, Euler also considered the same question for the hypothetical city illustrated here.

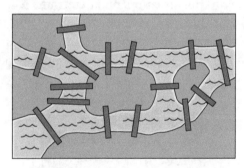

(a)   Draw a graph with the pieces of land as vertices and the bridges as edges.

(b)   Is it possible to start somewhere in the town and walk about the town crossing each bridge exactly once? The walk does not have to start and end at the same place. If it is possible, find such a path. If not, explain why not.

17. A meter reader wants to read the meter at each house in the three-block neighborhood pictured. He plans to park his truck at one of the street intersections and walk to each of the houses. All of the streets have houses on both sides, and he needs to travel down each street twice (once for each side). Draw a graph corresponding to the walking route he needs to cover. Determine whether or not the graph has an Eulerian circuit, and if so, give one. If it does not, explain why not.

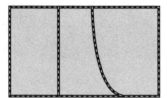

18. The sidewalks in a park are as illustrated in the figure. A city worker must sweep all of the sidewalks, and would like to cover each sidewalk exactly once. Draw a graph corresponding to this situation, and determine whether or not the graph has an Eulerian circuit, and if so, give one. If it does not, then explain why not. If the graph does not have an Eulerian circuit, determine whether it has an Eulerian path, and if it does, give one. If it does not, then explain why not.

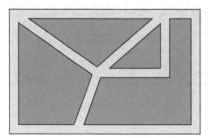

In Exercises 19 and 20, the parking meters along the streets of a city are indicated by dots in the given picture. A worker has the job of checking each of these meters and would like to find a route for which she will walk past each of the meters exactly once and not walk down streets without meters. Note that for the streets with meters on both sides, she will need to travel down the street twice. Draw a graph corresponding to the walking route she needs to cover. Determine whether or not the graph has an Eulerian circuit, and if so, give one. If it does not, then explain why not. If the graph does not have an Eulerian circuit, determine whether it has an Eulerian path, and if it does, give one. If it does not, then explain why not.

19.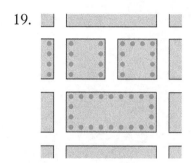

20.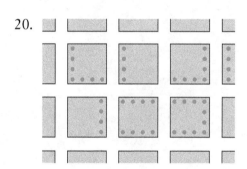

In Exercises 21 and 22, the floor plan for a house is shown.

(a) Draw a graph representing this floor plan by letting the rooms and the area outside the house be vertices and drawing an edge between any two vertices that have a door connecting them.

(b) Use the graph from part (a) to determine whether or not it is possible to start outside, go through each door of the house exactly once and end up outside again. If it is possible, give such a path. If it is not possible, explain why not.

21.

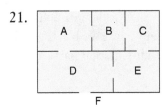

22.

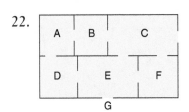

In Exercises 23–28, find an eulerization of the graph.

23.

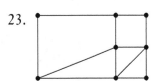

24.

25.

26.

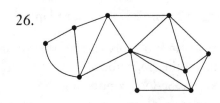

27.

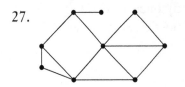

28.

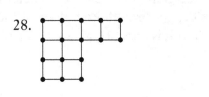

29. Suppose a garbage truck must pick up garbage in a neighborhood with streets laid out as in the graph. The workers pick up from both sides of the street at the same time, so ideally they would like to travel down each street as few times as possible. Find an eulerization of the graph that will allow the garbage truck to make a reasonably efficient Eulerian circuit through the neighborhood.

30. The layout of the garden at the Bonnefont Cloister in New York City is shown in the figure. Suppose a visitor to the garden wants to walk down each interior path at least once. Draw a graph corresponding to the walking route the visitor wants to cover. Find an eulerization of the graph that will allow the visitor to make an Eulerian circuit of the garden.

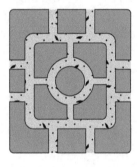

31. The sidewalks in a park are as shown in the figure. Suppose a woman wants to walk down each of the sidewalks at least once.

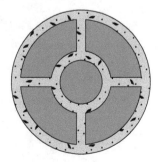

(a) Draw a graph corresponding to the walking route the woman wants to cover.

(b) Find an eulerization of the graph that will allow the woman to make an Eulerian circuit of the park.

(c) Assuming the lengths of the sidewalks in the picture are drawn to scale, does the eulerization you found in part (b) give the shortest tour of the park? If not, find one that gives a shorter tour.

32. Explain why every graph has an even number of odd vertices.

3.2 THE TRAVELING SALESMAN PROBLEM

In the last section, we were concerned with finding paths or circuits that cover every edge of a graph. We now look at the problem of visiting all of the vertices of a graph, without concern for whether or not all edges have been covered. This is a somewhat different way of thinking about graphs that comes up naturally in many real-life problems. For instance, a delivery person may be required to deliver packages to several locations. If we view the delivery locations as the vertices on a graph and the roads between the locations as the edges, then we see that the delivery person is only concerned with finding a path that touches each vertex and is not concerned with what edges are covered.

The problem of finding paths that visit vertices was studied by the Irish mathematician and astronomer Sir William Rowan Hamilton (1805–1865). Hamilton considered the problem of finding a circuit that goes through every vertex exactly once, returning to the starting vertex. These kinds of circuits are now named after him, and we have the following definition.

A circuit that begins at some vertex, goes through every other vertex exactly once, and returns to the starting vertex is called a **Hamiltonian circuit.**

A path that goes through each vertex exactly once, whether or not it returns to its staring vertex, is called a *Hamiltonian path*. However, we will be concerned primarily with Hamiltonian circuits in this section.

EXAMPLE 1 *Hamiltonian Circuits*

Find a Hamiltonian circuit, if possible, on the graphs in Figures 3.32 and 3.33.

(a)

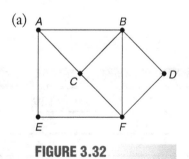

FIGURE 3.32

(b)

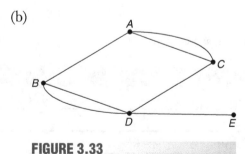

FIGURE 3.33

SOLUTION:

(a)   The circuit in Figure 3.34 is one possible answer. See if you can find other Hamiltonian circuits starting either at *A* or at some other vertex.

(b)   There are no Hamiltonian circuits on this graph because any circuit that goes through vertex *E* must go through vertex *D* at least twice.   ∎

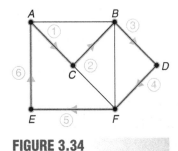

FIGURE 3.34

It would be nice to have a quick method for determining whether a graph has a Hamiltonian path or circuit, just as we are able to use Euler's Theorem to test whether a graph has Eulerian paths or circuits. Unfortunately, there is no known quick method for doing this. However, there are results that give us some information about particular graphs. One such result says that if a graph has at least three vertices and each vertex is connected to more than half of the other vertices in the graph, then the graph must have a Hamiltonian circuit.

Rather than studying the general problem of finding Hamiltonian circuits, we will consider a related problem, known as the traveling salesman problem, which has many practical applications. Before stating this problem precisely, we consider a specific instance of the traveling salesman problem.

Suppose a salesperson, starting in Philadelphia, must visit the cities of Cleveland, New York, and Pittsburgh on one car trip, and then return home to Philadelphia. If each city must be visited exactly once, what is the route that will minimize the distance of the entire trip? A map of the cities is shown in Figure 3.35, and the road mileage between these cities is listed in Table 3.1.

FIGURE 3.35

TABLE 3.1 Road Mileage

	Cleveland	New York	Philadelphia	Pittsburgh
Cleveland	—	473	413	129
New York	473	—	101	368
Philadelphia	413	101	—	288
Pittsburgh	129	368	288	—

We can use this information to construct a new kind of graph, a weighted graph. A **weighted graph** is a graph for which a number, called a **weight,** is assigned to each edge of the graph. In the graph modeling our current situation, each city will be represented by a vertex. There will be an edge between any two vertices and the edge will be assigned a weight equal to the road mileage between the two corresponding cities. The resulting weighted graph is shown in Figure 3.36.

Notice that the edges of the graph are not drawn to scale, nor are the cities laid out as they are on the map. We could draw this weighted graph as it appears on a map, but it is not necessary. Our salesman's route question can now be stated in terms of graph theory as follows: Find a Hamiltonian circuit that begins in Philadelphia for which the sum of the lengths of the edges along the circuit is a minimum. In this example, it is fairly easy to answer this question by simply checking all of the possible Hamiltonian circuits beginning in Philadelphia (there are only six in this case), and finding the one whose edges have the minimum total length. If you try this, you should find that the best route for the salesman is to follow the circuit shown in Figure 3.37, going first from Philadelphia to New York, next to Cleveland, then Pittsburgh, and finally back to Philadelphia, or the reverse route that goes first from Philadelphia to Pittsburgh, then to Cleveland, next to New York, and then back to Philadelphia.

The total mileage for either of these trips would be

$$101 + 473 + 129 + 288 = 991 \text{ miles.}$$

The problem we just looked at is only one example of the traveling salesman problem. The general problem applies to an entire class of graphs, known as complete weighted graphs. A graph is called **complete** if there is exactly one edge between every pair of vertices in the graph. A complete graph with six vertices is shown in Figure 3.38.

We can now state the traveling salesman problem.

The **traveling salesman problem** is the problem of finding a Hamiltonian circuit in a complete weighted graph for which the sum of the weights of the edges is a minimum.

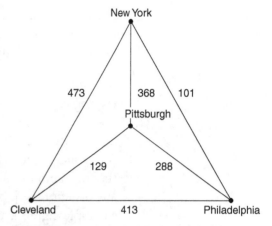

FIGURE 3.36

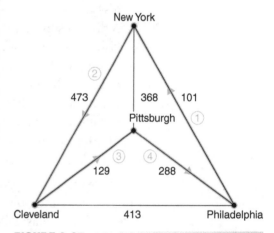

FIGURE 3.37

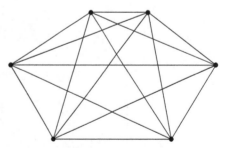

FIGURE 3.38 A complete graph with six vertices.

We refer to the sum of the weights of the edges of a circuit as the **weight of the circuit.** Thus, the solution to the traveling salesman problem is the circuit of minimum weight. It is important to note that in searching for a circuit of minimum weight, we can consider any particular vertex in the graph as the starting point for all of the circuits because every circuit can be considered as starting from any particular vertex lying on it. In most applications we will have a particular vertex in mind as the starting vertex.

The example of a salesman traveling to several cities that we looked at earlier is one instance of the traveling salesman problem. However, the traveling salesman problem also arises in other situations. A few such situations are as follows:

1. A water meter reader must find a walking route through a neighborhood for visiting each of the water meters he must read. In this case the vertices of the graph are the water meters, and the weighted edges are the routes between the meters with weight equal to the walking distance between each meter.

2. A technician must determine the order of a series of jobs on a machine so that the total amount of time required to go through all of the jobs is minimized. If any order of the jobs is possible, then we can think of the jobs as the vertices of a complete graph, and the time required to reset the machine from performing one job to performing the other as the weight of the edge joining these two jobs. For this situation, we are trying to minimize the total time rather than the total distance.

3. A district supervisor of several offices needs to find the most "efficient" route to visit each of the offices and return to her home office. In this case the vertices of the graphs are the offices. The weighted edges are the routes between the offices with weight equal to the distance or the driving time between the offices, depending on whether the supervisor is more interested in minimizing time or distance. Note that due to speed limits and traffic, minimizing time and distance may be very different problems.

We take this opportunity to emphasize that a weighted graph is an abstraction of the relevant features of a problem, and may not depict other features of the problem's setting. For instance, on the map in Figure 3.39(a), we see that Memphis lies between Little Rock and Nashville on Interstate 40. A complete weighted graph, based on mileage, for these three cities is shown in Figure 3.39(b). In traveling *directly* from

(a)   (b)

FIGURE 3.39 Comparing a map to a graph.

Little Rock to Nashville along Interstate 40, you would have to pass through Memphis, but this would not be indicated when you travel along the edge of the corresponding graph from Little Rock to Nashville. However, because a direct trip would not include a stop in Memphis, it is irrelevant that it is along the route and nothing is really lost in the translation to the graph.

How can we solve the traveling salesman problem for a given complete weighted graph? In other words, how do we find the Hamiltonian circuit of minimum weight? One method that is sure to work is to find the weight of every Hamiltonian circuit in the graph. Then the solution is the circuit or circuits (if there is a tie) whose weight is a minimum. This is how we found the solution for our salesman who was traveling from Philadelphia to Cleveland, New York, Pittsburgh, and back to Philadelphia. However, in that case the graph had only four vertices and there were only six possible circuits to check. To see why there were only six possibilities, note that after leaving Philadelphia, the salesman has three choices for the first city to visit. After making that choice, he has only two choices for the next city to visit, because he must go to one of the two remaining cities he has not visited. After making the choice for the second city to visit, the salesman has only one choice for the third city, because it is the only city left to visit before returning to Philadelphia. Therefore, the total number of Hamiltonian circuits beginning in Philadelphia is

$$3 \cdot 2 \cdot 1 = 6.$$

For a graph with n vertices the same argument applies, and we see that the total number of different Hamiltonian circuits starting at a particular vertex on the graph is given by

$$(n - 1) \cdot (n - 2) \cdots 3 \cdot 2 \cdot 1.$$

For any positive integer k, the number $k \cdot (k - 1) \cdots 3 \cdot 2 \cdot 1$ is called k *factorial* and is denoted by $k!$. Therefore, for a complete graph with n vertices, there are $(n - 1)!$ Hamiltonian circuits starting at a particular vertex. As the number of vertices, n, gets larger, the number $(n - 1)!$ gets large quickly, as we see in Table 3.2.

TABLE 3.2 Growth of $(n - 1)!$

n	$(n - 1)!$
4	6
8	5,040
12	39,916,800

With 12 vertices, it would still be possible to check all of the 39,916,800 possible Hamiltonian circuits fairly quickly on a computer. However, suppose we wanted to solve the traveling salesman problem on a complete weighted graph with 30 vertices. Then the number of Hamiltonian circuits to check is $29! \approx 10^{31}$.

Even if we could check these circuits at a rate of one billion per second on a computer, it would still take over 280 trillion years to check them all. For comparison, the estimated age of the universe is less than 20 billion years.

There are better methods than trying every possibility. For instance, we would no longer continue trying the different possibilities for a partial path that was doomed to exceed a value we have already found. Also, many problems have characteristics that allow us to rule out large numbers of paths as contenders for an optimal solution. In these kinds of problems, a traveling salesman problem with 30 vertices can be transformed into an easy computer exercise. Reducing the number of cases by deductive reasoning is a very important mathematical problem-solving technique that sometimes allows solutions to problems that would otherwise be overwhelming. An algorithm that takes a "reasonable" amount of time given the scale of the problem is called an *efficient algorithm*. Advances in computer hardware alone cannot keep pace with the number of vertices arising in practical applications. The attempt to find an efficient algorithm for solving the traveling salesman problem has taken on increased importance.

Although there is no known efficient algorithm that will solve the traveling salesman problem for every graph, there are methods that enable us to find a Hamiltonian circuit for which the weight is often fairly close to the minimum possible weight. These methods are called *approximate algorithms*. We will look at two relatively simple approximate algorithms that may yield fairly good answers for the traveling salesman problem. There are more advanced and better algorithms that, although rather difficult to carry out by hand, generally give excellent approximations when run on a computer.

The Nearest Neighbor Algorithm

The first approximate algorithm we will consider is called the **nearest neighbor algorithm,** and it is implemented as follows:

The Nearest Neighbor Algorithm for the Traveling Salesman Problem

(1)   Start at the vertex on which the circuit is supposed to begin and end.

(2)   At each vertex in the circuit, the next vertex in the circuit should be chosen to be the nearest vertex that has not yet been visited—in other words, the one connected to the current vertex by the edge of least weight. If there is a tie, choose any one of the nearest vertices. Mark the edges covered as you go along the path.

(3)   When all of the vertices have been visited, return to the starting vertex.

EXAMPLE 2 *Nearest Neighbor Algorithm*

Use the nearest neighbor algorithm to find an approximate solution to the traveling salesman problem for a circuit starting at vertex C on the graph in Figure 3.40.

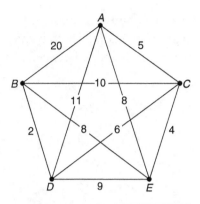

FIGURE 3.40

SOLUTION:

Starting at vertex C, the nearest neighbor is vertex E, so it should be the next vertex in the circuit, and we mark off our path as in Figure 3.41.

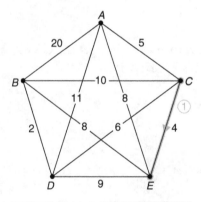

FIGURE 3.41

The nearest remaining vertices to E are vertices A and B. We may choose either one, so we choose vertex A and mark the edge we travel along as in Figure 3.42.

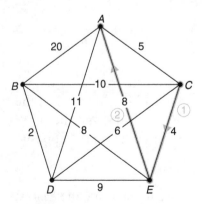

FIGURE 3.42

At this stage, we notice that vertex C is the nearest vertex to A, but we cannot go to vertex C because it has already been visited. Instead, we go to vertex D, which is the nearest vertex to A that has not yet been visited, and we mark the edge between A and D as in Figure 3.43.

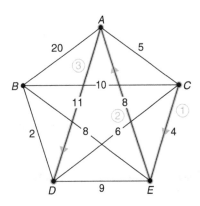

FIGURE 3.43

Continuing the algorithm, we go to vertex B and then return to C. We get the circuit in Figure 3.44. The weight of this circuit is $4 + 8 + 11 + 2 + 10 = 35$.

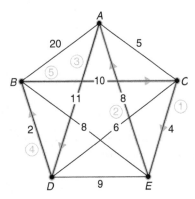

FIGURE 3.44

In Example 2, the nearest neighbor algorithm did not give the minimum weight circuit. In fact, by checking all circuits starting at C, we would find that one of the two circuits of minimum weight is the one shown in Figure 3.45, and the other circuit of minimum weight is the one given by traveling this circuit in the opposite direction.

These circuits both have weight 29. However, the circuit we found using the nearest neighbor algorithm was of much smaller weight than the maximum weight circuit starting at C, whose edges sum to 54. (See if you can find such a circuit.)

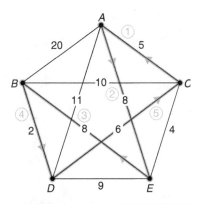

FIGURE 3.45

The Greedy Algorithm

We now look at another approximate algorithm for the traveling salesman problem, which we will call the **greedy algorithm.** It is sometimes called the *sorted edges algorithm*. It is similar to the nearest neighbor algorithm because we will choose shortest edges. However, instead of choosing the edge attached to the nearest vertex to our current vertex, we can choose the lowest weight edge available with a couple of restrictions that ensure we end up with a Hamiltonian circuit. More precisely, this algorithm can be stated as follows:

The Greedy Algorithm for the Traveling Salesman Problem

(1)   Begin by choosing the edge of least weight and marking the edge. If there is a tie between edges, choose any one.

(2)   At each stage, the next edge marked should be the unmarked edge of least weight unless it creates a circuit that does not visit every vertex or unless it results in three marked edges coming out of the same vertex in the graph. If there is a tie between edges, choose any one.

(3)   When the marked edges form a Hamiltonian circuit, the algorithm has been completed. The approximate solution is the marked circuit that begins at the starting vertex, travels along the marked edges in *either direction*, and returns to the starting vertex.

EXAMPLE 3   *Greedy Algorithm*

We will now use the greedy algorithm to find an approximate solution to the traveling salesman problem we considered in Example 2: finding a circuit starting at vertex *C* on the graph in Figure 3.40.

SOLUTION:

Starting with the graph in Figure 3.40, we see that the edge of least weight is the edge of weight 2 between vertices *B* and *D*. We mark this edge as in Figure 3.46.

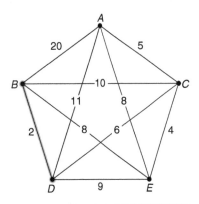

FIGURE 3.46

The unmarked edge of least weight at this stage is the edge between vertices C and E of weight 4. We mark it next as in Figure 3.47.

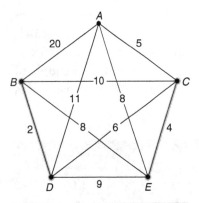

FIGURE 3.47

Next, we mark the edge of weight 5 between vertices A and C as in Figure 3.48.

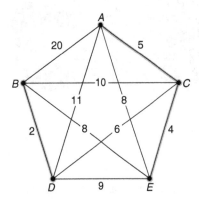

FIGURE 3.48

At this point, the unmarked edge of the least weight is the edge between vertices C and D of weight 6. However, we cannot mark this edge, because doing so would result in three marked edges coming out of vertex C. Therefore, we move on to consider the two edges of weight 8. In the case of a tie, we would normally be able to choose either one. However, in this case, marking the edge between vertices A and E would create a circuit. Therefore, we mark the edge between vertices B and E as in Figure 3.49.

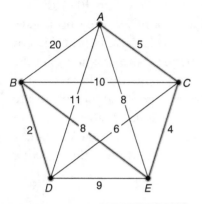

FIGURE 3.49

Continuing the greedy algorithm, the next edge that we will mark is the edge of weight 11 between vertices A and D. The last edge we add using the greedy algorithm is always the only unmarked edge left that will close the circuit, as shown in Figure 3.50.

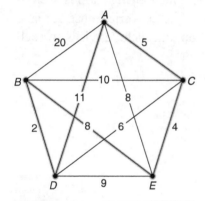

FIGURE 3.50

We are looking for a circuit starting and ending at vertex C. Starting at vertex C and following the edges marked in Figure 3.50, we get either the circuit shown in Figure 3.51 or the same circuit in reverse. Each of these circuits is an approximate solution given by the greedy algorithm, and each of them has weight $4 + 8 + 2 + 11 + 5 = 30$.

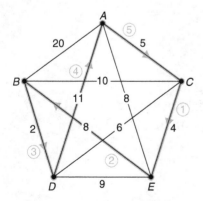

FIGURE 3.51

In Examples 2 and 3 we found approximate solutions to the same traveling salesman problem using two different algorithms. The nearest neighbor algorithm gave a circuit of weight 35 and the greedy algorithm gave a circuit of weight 30. Although neither of these gave the optimal circuit, which has weight 29, they both gave solutions much better than the worst possible circuit, which has weight 54. For some graphs, the nearest neighbor algorithm gives a better approximation than the greedy algorithm, and sometimes they give the same approximation. In fact, in the problem we looked at in Examples 2 and 3, the nearest neighbor algorithm would have given the same circuit as the greedy algorithm if we had made a different choice in the case of the tied edges. With some luck, either of these approximate algorithms may give the optimal circuit and, therefore, the solution to the traveling salesman problem.

EXAMPLE 4 *Delivery Truck Route*

A copying service must make five deliveries downtown at the locations labeled on the map in Figure 3.52 with the letters *A*, *B*, *C*, *D*, and *E*. The copying shop is located at the spot labeled *S*. All of the blocks are of equal size. The deliveries must be made along a route that begins and ends at the copying shop and includes visits to each delivery location exactly once.

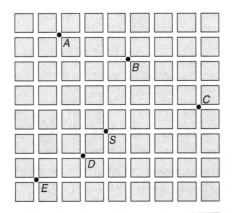

FIGURE 3.52 Map of the town.

(a) Draw a complete weighted graph corresponding to the problem.
(b) Find a route and its length using the nearest neighbor algorithm.
(c) Find a route and its length using the greedy algorithm.

SOLUTION:

(a) The vertices of the complete weighted graph corresponding to this situation are the copying shop and the five delivery locations. The weight of an edge between two vertices is the distance between the corresponding locations on the map, measured in blocks that must be traveled along the roads. We will always take the shortest distance possible. For instance, the weight of the edge between vertex E and S is 5, because we must travel three blocks east and two blocks north to get from E to S. Finding all of the weights in this way, we get the graph in Figure 3.53.

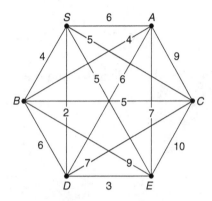

FIGURE 3.53

Notice that all of the relevant geometry from the original problem has been incorporated in the weights for this complete graph.

(b) When we apply the nearest neighbor algorithm for a circuit starting at the copying shop S, we do not run into any step where we have a tie, and the resulting circuit is drawn in Figure 3.54. The length of the entire route is $2 + 3 + 7 + 4 + 5 + 5 = 26$ blocks.

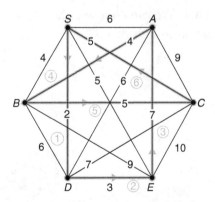

FIGURE 3.54

A driving route corresponding to the circuit we found with the nearest neighbor algorithm is drawn in Figure 3.55. Note, however, that this is just one of many routes going through the points in the same order and covering the same distance. For instance, in going from point S to point D, the route could have gone one block west and then one block south, rather than as it was drawn going south first and then west.

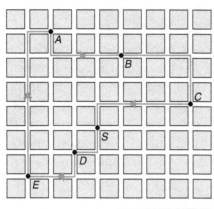

FIGURE 3.55

(c)   With the greedy algorithm, we begin by marking the edge of length 2 and then the edge of length 3, as in Figure 3.56.

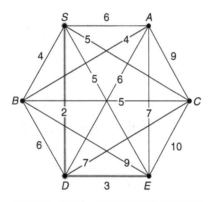

FIGURE 3.56

Now we must choose one of the edges of length 4. We can choose either, and in this graph, we then choose the other one next, as in Figure 3.57.

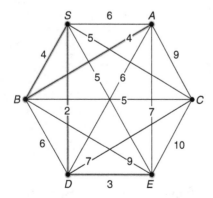

FIGURE 3.57

We see that we cannot mark any of the edges of length 5, 6, or 7, because each one will create a circuit or result in three marked edges coming out of the same vertex. We also cannot mark the edge of weight 9 between vertices B and E, so we mark the edge of weight 9 between vertices A and C, as in Figure 3.58.

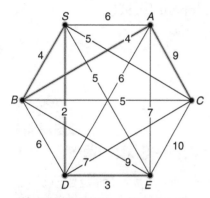

FIGURE 3.58

We now complete the algorithm by marking the edge between vertices C and E, and we get the circuit in Figure 3.59.

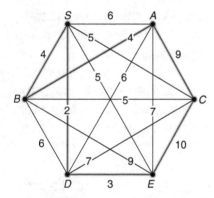

FIGURE 3.59

Our circuit must start and end at vertex S, so the greedy algorithm has yielded the circuit in Figure 3.60 or the reverse circuit.

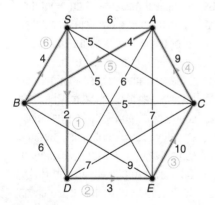

FIGURE 3.60

The lengths of the corresponding routes are both $2 + 3 + 10 + 9 + 4 + 4 = 32$ blocks. We see that, in this case, the nearest neighbor algorithm gave a better approximation to the traveling salesman problem than did the greedy algorithm. In fact, the nearest neighbor algorithm gave the optimal route. ■

branching OUT

3.2

WHY EFFICIENT ALGORITHMS ARE IMPORTANT

Unlike questions in other areas, questions about specific graphs, such as finding the Hamiltonian circuit of minimal weight on a complete weighted graph, are always theoretically solvable. Because the graph is finite, we need only check all possibilities, for instance all possible Hamiltonian circuits. The only issue is the practical one of time. Thirty vertices is fairly small for a typical application. We live in a world where computers can execute a billion instructions per second. Yet even if computers were a thousand times faster than this, all of the computers in the world working together could not carry out a brute force check of the weights of all of the Hamiltonian circuits in a complete weighted graph with 30 vertices within our lifetimes.

A great algorithm would process the input data, such as the weights of the edges of a complete weighted graph, and determine the answer, such as the Hamiltonian circuit of minimal weight, in a time proportional to the size of the input. Thus, doubling the amount of data would approximately double the time taken to determine the answer, tripling the input would triple the time, and so forth. Such an algorithm is said to work in *linear time*. Translating passages from one language to another approximates the idea of a linear time algorithm. A more mathematical example would be adding two numbers, where the size of the input is the number of digits of the numbers. For more complex problems, it is too much to hope for a linear time algorithm. For instance, the length of time required

to multiply two numbers goes up by a factor of four when the numbers double in length, by a factor of nine when the numbers triple in length, and so forth. Standard multiplication is thus an example of a *quadratic time* algorithm, with computing times growing in a manner similar to that of the square of the size of the input, or n^2, where n is the size of the input. More generally, we could speak of a *polynomial time* algorithm, with computing times growing in a manner similar to n^k, where n is the size of the input and k is a positive integer. Whether there exists a polynomial time algorithm for solving the traveling salesman problem is one of the most famous unsolved problems in all of mathematics.

Whether an algorithm is linear, quadratic, polynomial, or worse is a statement about the relative long-term efficiency of the algorithm. However, this is only half of the story. If an algorithm takes a million years to solve even a small-scale problem, it does not matter much whether the algorithm is linear. Thus, a practical measure of an algorithm is how long it takes to solve a problem. The speed of the computer or computers is as much a factor of whether an algorithm is practical or not as the actual steps of the algorithm. Thus, an algorithm that is currently viewed as impractical may become practical as technology improves. Another measure of the practicality of an algorithm is how large a problem it can solve in a "reasonable" amount of time. In 1998, David Applegate, Robert Bixby, and William Cook at the Center for Research on

Parallel Computation at Rice University, along with Vasek Chvátal (Rutgers University), solved a traveling salesman problem for 13,509 U.S. cities with populations of more than 500 people. The solution is shown below. Using several computers for processing, it took over 3 months to find the solution. To date, the largest solved traveling salesman problem was on a graph with 85,900 vertices. Completed in April 2006, it took over 136 years of total computing time to arrive at the solution. While these are two very large solved cases of the traveling salesman problem, there is not currently an algorithm that would be guaranteed to solve every possible 100 vertex problem in a reasonable amount of time on today's computers, and it is not believed that there will ever be such an algorithm.

In practical settings, any improvement of the current practice translates into real monetary savings. Thus, finding approximate solutions to problems such as the traveling salesman problem has immense practical importance. In many, if not most, practical problems, the weight of the edge between two vertices cannot exceed the combined weight obtained by passing through one or more intermediate vertices. In such instances, a simple variation of Prim's algorithm for finding the minimal weight network, treated in the next section, easily finds a Hamiltonian circuit that is no more than twice the weight of the minimal weight circuit. A modification of this idea can decrease this factor from 2 to 1.5. More sophisticated methods, such as the Lin–Kernighan method, yield circuits that are usually within a few percent of the optimal solution. Such techniques take only a few minutes when applied to a problem with 1000 vertices. Even a problem with a million vertices can usually be solved to within a few percent of the optimum in several hours on a fast computer. In the context of a massive network such as an oil pipeline or the circuit layout used in manufacturing millions of computer chips, saving even a few percent can represent millions of dollars. It is far from unusual for applications of graph theory to yield savings of 25% or more. Thus, the search for even slight improvements in existing algorithms is an important and active area of research in mathematics and computer science.

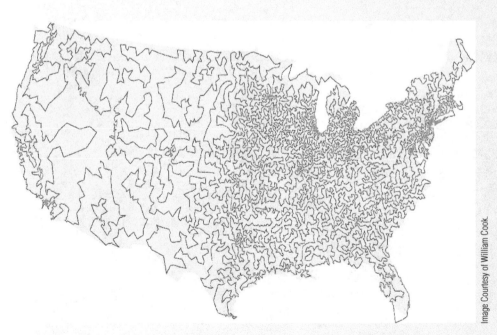

Image Courtesy of William Cook.

Solution of a traveling salesman problem with 13,509 cities in the United States.

EXERCISES / FOR SECTION 3.2

In Exercises 1–6, find a Hamiltonian circuit, if possible, in the given graph. If it is not possible to find a Hamiltonian circuit, explain why not.

1.

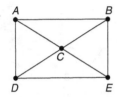

2.

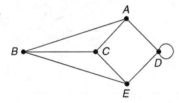

3.

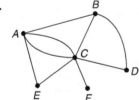

4.

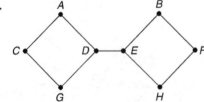

5.

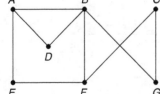

6.

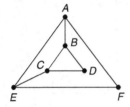

In Exercises 7–14, find an approximate solution to the traveling salesman problem for a circuit starting at vertex A on the given weighted graph and find the weight of the circuit using (a) the nearest neighbor algorithm, (b) the greedy algorithm.

7. (a)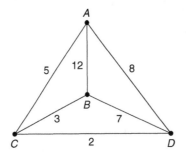

(b)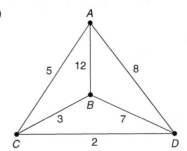

8. (a)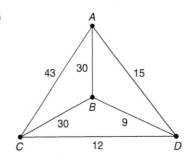

(b)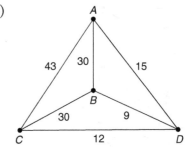

9. (a)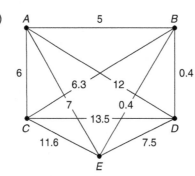

(b)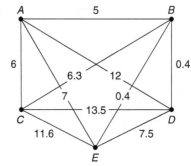

10. (a)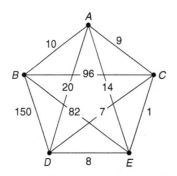

(b)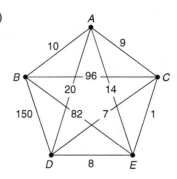

11. (a) 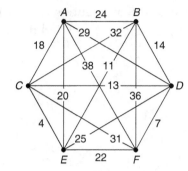   (b)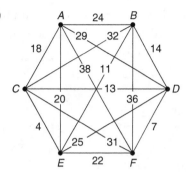

12. (a) 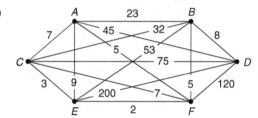   (b)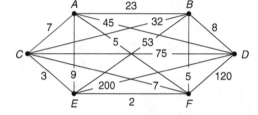

13. (a) 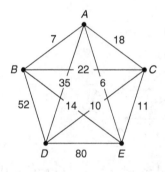   (b)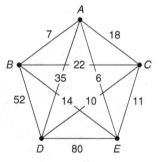

14. (a) 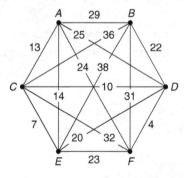   (b)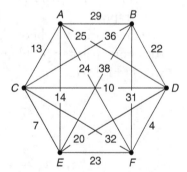

15. This problem will explore how different algorithms play out with the same weighted graph.
    (a) Use the nearest neighbor algorithm to find an approximate solution to the traveling salesman problem for a circuit starting at vertex C, and find the weight of this circuit.

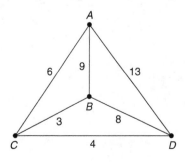

(b)   Use the greedy algorithm to find an approximate solution to the traveling salesman problem for a circuit starting at vertex *C*, and find the weight of this circuit.

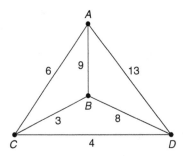

(c)   By checking all possible circuits, find an exact solution of the traveling salesman problem for a circuit starting at vertex *C*.

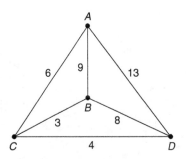

16.   This problem will explore how different algorithms play out with the same weighted graph.

(a)   Use the nearest neighbor algorithm to find an approximate solution to the traveling salesman problem for a circuit starting at vertex *D*, and find the weight of this circuit.

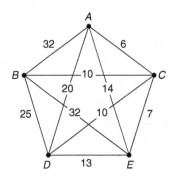

(b) Use the greedy algorithm to find an approximate solution to the traveling salesman problem for a circuit starting at vertex D, and find the weight of this circuit.

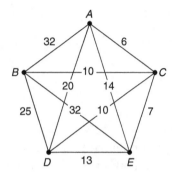

(c) It turns out that neither the nearest neighbor algorithm nor the greedy algorithm gives a circuit of minimum total weight starting at vertex D. Find some circuit starting at D of total weight smaller than those of the circuits you found in parts (a) and (b).

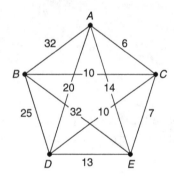

17. A person starting in Wichita must visit Kansas City, Omaha, and St. Louis, and then return home to Wichita in one car trip. A map including these cities and the road mileage between them is shown.

	Kansas City	Omaha	St. Louis	Wichita
Kansas City	—	201	256	190
Omaha	201	—	449	298
St. Louis	256	449	—	447
Wichita	190	298	447	—

(a) Draw a weighted graph corresponding to this situation.

(b) Use the nearest neighbor algorithm to find an approximate solution to the traveling salesman problem for a circuit starting at Wichita, and find the length of this circuit.

(c) Use the greedy algorithm to find an approximate solution to the traveling salesman problem for a circuit starting at Wichita, and find the length of this circuit.

(d) By checking all possible routes, find the route that will minimize the distance of the entire trip, and find the total length of this route. How does the length of this route compare to the lengths of the routes you found in parts (b) and (c)?

18. In January 2012, the one-way airline fares on American Airlines were those listed in the following table. Suppose a company president, who lived in Nashville, needed to travel to offices in Los Angeles, New York, Dallas, and Washington, D.C., and then return to Nashville in one trip.

	Los Angeles	Nashville	New York City	Dallas	Washington, D.C.
Los Angeles	—	$191	$139	$209	$139
Nashville	$191	—	$216	$148	$200
New York City	$139	$216	—	$223	$ 69
Dallas	$209	$148	$223	—	$196
Washington, D.C.	$139	$200	$ 69	$196	—

(a) Draw a weighted graph corresponding to this situation.

(b) Use the nearest neighbor algorithm to find a travel route and find the total cost of traveling this route.

(c) Use the greedy algorithm to find a travel route and find the total cost of traveling this route.

19. A tourist visiting New York City wants to visit the Empire State Building, the Statue of Liberty, the Brooklyn Bridge, and Times Square. She will begin and end her tour at Grand Central Station and travel to each of her stops by subway. The estimated travel time, in minutes, on the New York subway between each of the stops is given in the table.

New York City subway map.

	Grand Central Station	Empire State Building	Statue of Liberty	Brooklyn Bridge	Times Square
Grand Central Station	—	15	30	5	5
Empire State Building	15	—	30	20	5
Statue of Liberty	30	30	—	30	20
Brooklyn Bridge	5	20	30	—	15
Times Square	5	5	20	15	—

(a) Draw a weighted graph corresponding to this situation.

(b) Use the nearest neighbor algorithm to find a route, and find the estimated time required to travel this route.

(c) Use the greedy algorithm to find a route, and find the estimated time required to travel this route.

20. A pizza delivery man must deliver pizzas to the four locations on the following map labeled with the letters *A*, *B*, *C*, and *D*. The pizza shop is labeled with the letter *P*. All of the blocks are of equal size. The delivery man needs to travel a route that begins and ends at the pizza shop and includes a visit to each delivery location exactly once.

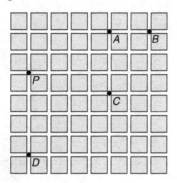

(a) Draw a complete weighted graph corresponding to the problem.

(b) Find a route and its length using the nearest neighbor algorithm.

(c) Find a route and its length using the greedy algorithm.

21. A truck driver must deliver newspapers to each of the five newsstands sitting on street corners in a small town. The truck's route will begin and end at the newspaper's main office. The map shows the location of the main office labeled with the letter M and the locations of the newsstands labeled with the letters A, B, C, D, and E. All blocks are of equal size. The truck needs to travel a route that begins and ends at the main office and includes a visit to each newspaper stand exactly once.

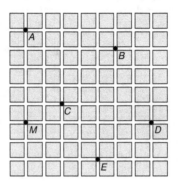

(a) Draw a complete weighted graph corresponding to the problem.

(b) Find a route and its length using the nearest neighbor algorithm.

(c) Find a route and its length using the greedy algorithm.

For Exercises 22 and 23, consider the following: Obtaining viewing times on the large telescopes at the major observatories is often a highly competitive process. Because such opportunities are rationed out, it is desirable to minimize the time spent moving the telescope into position. Furthermore, minimizing the telescope's movement may prolong its life. Thus, it may be reasonable to minimize the total angular change as the telescope is adjusted to view different objects.

22. Suppose an astronomer wants to view five stars, and the angles (in degrees) between the stars and between the stars and the base position of the telescope are given in the table.

	Base	Polaris	Betelgeuse	Rigel	Vega	Alphard
Base	—	8.0	80.8	96.6	54.7	91.7
Polaris	8.0	—	82.1	96.8	51.6	98.8
Betelgeuse	80.8	82.1	—	17.2	132.8	55.3
Rigel	96.6	96.8	17.2	—	144.1	61.3
Vega	54.7	51.6	132.8	144.1	—	131.4
Alphard	91.7	98.8	55.3	61.3	131.4	—

(a) Draw a weighted graph corresponding to this situation.

(b) Use the nearest neighbor algorithm to find an approximate solution to the traveling salesman problem for a circuit starting at the base position, and find the total angular change when traveling this circuit.

(c) Use the greedy algorithm to find an approximate solution to the traveling salesman problem for a circuit starting at the base position, and find the total angular change when traveling this circuit.

23. Suppose an astronomer wants to view seven comets, and the angles (in degrees) between the comets and between the comets and the base position of the telescope are given in the table. *Without drawing the corresponding graph*, use the nearest neighbor algorithm to find an approximate solution to the traveling salesman problem for a circuit starting at the base position, and find the total angular change when traveling this circuit.

	Base	Barnard 3	du Toit–Hartley	Gunn	Harrington–Wilson	Neujmin 2	Shoemaker–Levy 7	Swift
Base	—	46.4	53.7	85.0	71.1	76.9	61.7	119.1
Barnard 3	46.4	—	59.4	39.2	93.4	55.8	20.0	72.8
du Toit–Hartley	53.7	59.4	—	73.0	34.8	114.8	78.2	102.4
Gunn	85.0	39.2	73.0	—	104.3	61.9	32.3	35.8
Harrington–Wilson	71.1	93.4	34.8	104.3	—	146.3	112.8	123.4
Neujmin 2	76.9	55.8	114.8	61.9	146.3	—	39.1	64.7
Shoemaker–Levy 7	61.7	20.0	78.2	32.3	112.8	39.1	—	58.5
Swift	119.1	72.8	102.4	35.8	123.4	64.7	58.5	—

24. A rabid Chicago Cubs fan decides to spend his summer vacation visiting each of the stadiums of the National Baseball League. The mileage between each of the National League cities is given in the chart. He plans to begin and end his

trip in Chicago and pass through each city exactly once. *Without drawing the corresponding graph*, use the nearest neighbor algorithm to find a route.

	Atlanta	Chicago	Cincinnati	Denver	Houston	Los Angeles	Miami	Milwaukee	New York	Philadelphia	Pittsburgh	Phoenix	St. Louis	San Diego	San Francisco	Washington, D.C.
Atlanta	—	674	440	1398	789	2182	655	761	841	741	687	1793	541	2126	2496	638
Chicago	674	—	287	996	1067	2054	1329	90	802	738	452	1713	289	2064	2142	700
Cincinnati	440	287	—	1173	1029	2179	1095	374	647	567	295	1804	340	2155	2362	510
Denver	1398	996	1173	—	1019	1059	2037	1029	1771	1691	1411	792	857	1108	1235	1658
Houston	789	1067	1029	1019	—	1538	1190	1142	1608	1508	1313	1149	839	1482	1912	1411
Los Angeles	2182	2054	2179	1059	1538	—	2687	2087	2786	2706	2426	389	1845	124	379	2669
Miami	655	1329	1095	2037	1190	2687	—	1416	1308	1208	1200	2390	1196	2672	3053	1055
Milwaukee	761	90	374	1029	1142	2087	1416	—	889	825	539	1751	363	2102	2175	787
New York	841	802	647	1771	1608	2786	1308	915	—	101	368	2411	948	2762	2934	226
Philadelphia	741	738	567	1691	1508	2706	1208	889	101	—	288	2331	868	2682	2866	137
Pittsburgh	687	452	295	1411	1313	2426	1200	825	368	288	—	2051	588	2402	2578	245
Phoenix	1793	1713	1804	792	1149	389	2390	539	2411	2331	2051	—	1470	358	763	2300
St. Louis	541	289	340	857	839	1845	1196	1751	948	868	588	1470	—	1821	2089	816
San Diego	2126	2064	2155	1108	1482	124	2672	363	2762	2682	2402	358	1821	—	504	2691
San Francisco	2496	2142	2362	1235	1912	379	3053	2102	2934	2866	2578	763	2089	504	—	2815
Washington, D.C.	638	700	510	1658	1411	2669	1055	787	226	137	245	2300	816	2691	2815	—

Exercises 25 and 26 illustrate the variety of possible applications for the following: When the vertices of a graph are located in a plane and the weight of an edge is simply the distance between the two vertices, we can use the distance formula to find the length of the edges. The distance between points with coordinates (x_1, y_1) and (x_2, y_2) is

$$\sqrt{(x_2 - x_1)^2 + (y_2 - y_1)^2}.$$

25. A drill must drill holes at four points on a circuit board. The points have coordinates $(1, 0)$, $(4, 1)$, $(0, 2)$, and $(5, 4)$, measured in centimeters. The drill begins at the point $(1, 0)$ and must complete a round-trip through each of the other three locations. (The round-trip is so it can drill the next board in the production line.)

    (a) Draw a weighted graph corresponding to this situation.

    (b) Use the nearest neighbor algorithm to find an order in which to drill the holes, and find the length the drill must be moved as it follows this order.

    (c) Use the greedy algorithm to find an order in which to drill the holes, and find the length the drill must be moved as it follows this order.

26. A golfer hits three balls from the point (0, 0) that come to rest at the points (97, –15) (89, 2), and (95, 11), measured in yards. The golfer wishes to retrieve the balls and go back to the starting point to hit them again.
    (a) Draw a weighted graph corresponding to this situation.

    (b) Use the nearest neighbor algorithm to find an order in which the golfer should retrieve the balls and the length of the corresponding walking route.

    (c) Use the greedy algorithm to find an order in which the golfer should retrieve the balls and the length of the corresponding walking route.

27. Sir William Rowan Hamilton, after whom Hamiltonian circuits were named, marketed a game called Icosian consisting of a wooden dodecahedron (the regular solid with 12 pentagonal faces as shown in Figure 7.47) with pegs attached to each corner and a string. The corners of the dodecahedron were labeled with the names of different cities of the world. The object of the game was to travel from city to city along the edges without going through any city twice and then return to the starting point. The circuit was marked off using the string. When a dodecahedron

is flattened out it looks like the graph shown, where the edges and vertices of the graph correspond to the edges and corners of the dodecahedron. Therefore, finding a Hamiltonian circuit on this graph is equivalent to the object of Hamilton's game. Find a Hamiltonian circuit on this graph starting at vertex D.

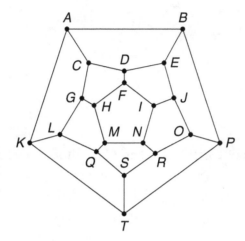

Exercises 28 and 29 consider another method for finding an approximate solution to the traveling salesman problem. With this method the nearest neighbor algorithm is applied repeatedly, using each of the vertices of the graph as the starting vertex in the algorithm. From these circuits, the one with the shortest length can be rewritten so that the starting vertex is the one specified by the particular problem, giving an approximate solution. Apply this method to find an approximate solution to the traveling salesman problem starting at vertex B, and give the weight of this circuit.

28.

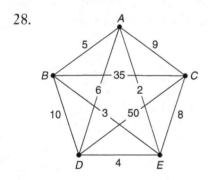

29.

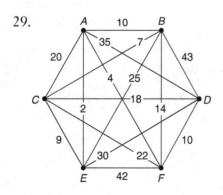

30. Draw an example of a graph that has an Eulerian circuit but does not have a Hamiltonian circuit.

31. Draw an example of a graph that has a Hamiltonian circuit but does not have an Eulerian circuit.

32. Draw an example of a complete weighted graph with at least four vertices for which the nearest neighbor algorithm starting at some vertex gives the worst possible Hamiltonian circuit. In other words, all other Hamiltonian circuits, except for the same circuit in reverse order, should have smaller weight.

33. Draw an example of a complete weighted graph with at least four vertices for which the greedy algorithm gives the worst possible Hamiltonian circuit. In other words, all other Hamiltonian circuits, except for the same circuit in reverse order, should have smaller weight.

34. Suppose you want to find the length of the longest Hamiltonian circuit in a complete weighted graph. How can the idea of the nearest neighbor algorithm be adapted to give an algorithm for finding an approximate solution to this problem?

3.3   EFFICIENT NETWORKING: MINIMAL SPANNING TREES

There are many instances when we want to connect several points, but when a direct connection between each pair of points is not necessary. Consider, for instance, eight computers in a building that need to be networked—that is, connected to one another by cable. Suppose it is most convenient to install the cable alongside the existing wiring in the building, and the eight computers and current wiring are laid out as shown in Figure 3.61, where the boxes indicate the computers, the edges indicate the existing wiring, and the numbers along each edge indicate the length of the edge in yards.

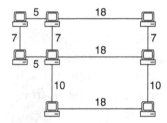

FIGURE 3.61 Computers and existing wiring.

One way to network all of the computers is shown in Figure 3.62, where the lines marked in blue indicate where the new cable will be installed.

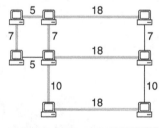

FIGURE 3.62

Notice that in Figure 3.62, each computer is connected to every other computer by cable, although some of the connections are not direct because they go through one or more other computers. There is no need for new cabling along any of the unused edges in Figure 3.62 because all computers are now networked. However, we might wonder if we networked the computers using the least amount of cable possible. Adding up all of the edges with cable in Figure 3.62, we see that we would need $5 + 18 + 7 + 7 + 18 + 10 + 18 = 83$ yards of cable. It turns out that we could have done better than this if we had installed the cable as in Figure 3.63.

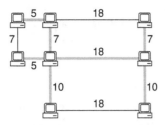

FIGURE 3.63

The total length of the cable in Figure 3.63 is 5 + 7 + 7 + 5 + 18 + 10 + 10 = 62 yards, a significant improvement over the 83 yards used in Figure 3.62. Using a method we will learn later in this section, we can show that this is the least amount of cable that can be used to form the network.

In terms of graph theory, a networking problem asks us to efficiently connect all of the vertices of a given graph. The solution will always be a graph within the original graph that is connected, contains all of the vertices, and contains no circuits. We will usually be interested in minimizing the weight of such a network. Some terms from graph theory are needed to help us to describe networking problems.

A graph lying within the original graph is called a subgraph. A **subgraph** is any collection of edges and vertices from the original graph that themselves form a graph. The graph colored blue within the graph in Figure 3.64 is an example of a subgraph.

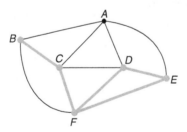

FIGURE 3.64 A subgraph.

We have the following important definitions:

> Any graph that is connected and contains no circuits is called a **tree**. A subgraph that is a tree containing all of the vertices of the graph is called a **spanning tree** for the graph.

In Figure 3.65(a), we see a subgraph that is a spanning tree. The subgraph in Figure 3.65(b) is not a tree because it contains a circuit and therefore it is not a spanning tree. The subgraph in Figure 3.65(c) is not a tree because it is not connected, and therefore it cannot be a spanning tree. The subgraph in Figure 3.65(d) is a tree

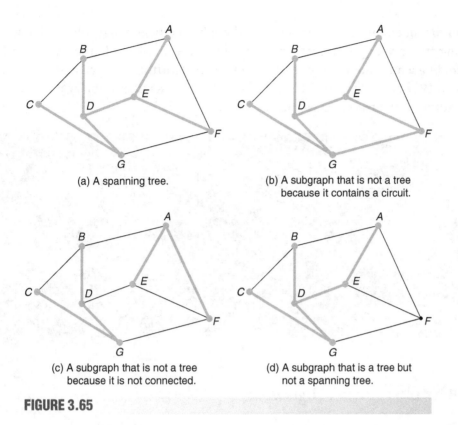

(a) A spanning tree.

(b) A subgraph that is not a tree
because it contains a circuit.

(c) A subgraph that is not a tree
because it is not connected.

(d) A subgraph that is a tree but
not a spanning tree.

FIGURE 3.65

because it is connected and contains no circuits, but it is not a spanning tree because it does not contain the vertex F.

Because spanning trees give a connection between every pair of vertices in the graph, they provide a network connecting all of the vertices of the graph. Because there are no circuits in a spanning tree, there are no unnecessary edges. In many applications, the graph will be weighted, and we will be interested in finding a minimum spanning tree, which is defined as follows:

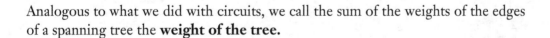

A **minimum spanning tree** for a weighted graph is a spanning tree for which the sum of the weights of the edges is a minimum.

Analogous to what we did with circuits, we call the sum of the weights of the edges of a spanning tree the **weight of the tree.**

Prim's Algorithm

The problem of finding a minimum spanning tree for a weighted graph is quite similar to solving the traveling salesman problem, where we are looking for the minimum

Hamiltonian circuit. Surprisingly, even though no efficient algorithm is known for solving the traveling salesman problem, there are simple and efficient algorithms for finding a minimum spanning tree. One such algorithm was presented by Robert Prim in 1957 and is known as **Prim's Algorithm.** It is similar to the nearest neighbor algorithm for the traveling salesman problem.

Prim's Algorithm for Finding a Minimum Spanning Tree

(1)   Start at any vertex in the graph. This vertex will be the starting tree for the algorithm.

(2)   Find the nearest vertex to the current tree that is not already contained in the current tree. In other words, from among those vertices not in the current tree, find the one that may be connected to a vertex in the current tree by an edge of the least weight. Connect this vertex to the current tree by the edge of least weight connecting them. If there is a tie between nearest vertices, choose any one of them.

(3)   Continue the procedure in step 2 until all of the vertices are contained in the tree.

EXAMPLE 1 *Prim's Algorithm*

Using Prim's algorithm, find a minimum spanning tree for the weighted graph in Figure 3.66 and give the weight of the tree.

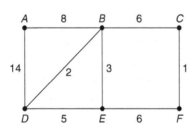

FIGURE 3.66

SOLUTION:

We can start at any vertex we wish, so we start with vertex *B*. This is the start of our minimum spanning tree. The nearest vertex to *B* is vertex *D*, so we choose the edge between *B* and *D* and mark it as in Figure 3.67.

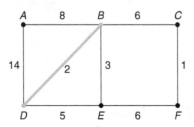

FIGURE 3.67

The nearest vertex to the current tree is vertex E. We connect it to vertex B and get the tree in Figure 3.68.

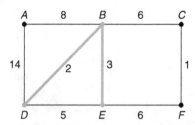

FIGURE 3.68

The nearest vertices to the current tree that are not already contained in it are vertices C and F. Because it is a tie, we can connect either one. Connecting vertices C and B results in the tree in Figure 3.69.

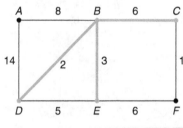

FIGURE 3.69

The nearest vertex to this tree is vertex F, and connecting it to vertex C we get the tree in Figure 3.70.

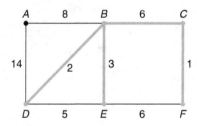

FIGURE 3.70

Finally, we connect the last vertex A to vertex B and we have found the minimum spanning tree shown in Figure 3.71.

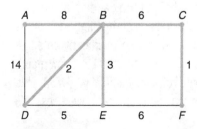

FIGURE 3.71

The weight of this spanning tree is the sum of the weights of the edges: 8 + 6 + 2 + 3 + 1 = 20. ■

branching OUT

STEINER POINTS: IMPROVING ON THE MINIMAL SPANNING TREE

In planning communication networks, several points, such as cities, must be connected. It is natural to think of these points as the vertices in a graph and of the straight lines between the points as the edges. However, when trying to find a minimal network connecting these points, it is often desirable to add other vertices and edges.

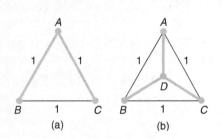

(c) Minimal network allowing Steiner points.

This may reduce the length of the network. For instance, suppose we want to connect three cities lying at the corners, *A, B,* and *C,* of an equilateral triangle. If the sides of the triangle have length 1, then one minimal spanning tree of the graph with vertices *A, B,* and *C,* and edges formed by the triangle is shown in figure (a). We see that the minimal spanning tree has length 2. However, if we augment the graph by adding a vertex at the center of the triangle and joining each of the original vertices to this new vertex by an edge, then the minimal spanning tree of this new graph, shown in figure (b), has length $\sqrt{3} \approx 1.732$. This is the best we can do for connecting vertices *A, B,* and *C.* When our vertices are points in the plane, added vertices are called *Steiner points.* Allowing the addition of Steiner points gives many more possibilities for a minimal network. It lets us reduce the total length, but makes the problem of being sure we have the best solution more difficult. Using the method of Ernest J. Cockayne and Denton E. Hewgill of the University of Victoria, the minimal network connecting 29 cities in North America and allowing Steiner points was found and the total length is about 7400 miles. It is shown in figure (c).

By comparison, the minimal spanning tree of the complete graph formed by the 29 cities and the straight lines connecting them to one another, which is shown in figure (d), has a total length of about 7600 miles.

Although many properties of Steiner points are known, such as that three edges meet at 120° angles at any Steiner point in a minimal configuration, there is no known efficient algorithm in the spirit of Prim's algorithm for finding a minimal network that allows Steiner points. Mathematicians Frank Hwang and Ding-Zhu Du proved in 1990 that in going from the minimal spanning tree to the minimal network with Steiner points allowed, the weight of the network is reduced by at most $(2 - \sqrt{3})/2 \approx 13.4\%$ This was the savings obtained in the example of figure (b). Although the minimal spanning tree is not a bad approximation to the minimal length network including Steiner points, you should remember that with enormous networks, a few percent savings may translate to millions of dollars.

An interesting alternative to computational algorithms that sometimes improves on the minimal spanning tree is to place pegs on a surface to mark the local of vertices and to stretch soap bubbles across these pegs. Soap bubbles are known to form in ways that tend to minimize area, which in this case is equivalent to the length of the network. The resulting soap bubble network has Steiner points, but it is not necessarily the minimal one allowing Steiner points.

(d) Minimal spanning tree using only the original vertices.

Image © Volina, 2012. Shutterstock, Inc.

A graph may have only one minimal spanning tree or several minimum spanning trees. Which spanning tree we find using Prim's algorithm can depend on the choices we make when we have to decide between the two or more nearest vertices in the case of a tie. For example, another minimum spanning tree for the graph in Example 1 is shown in Figure 3.72.

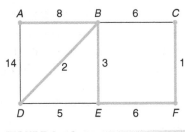

FIGURE 3.72

However, when the edges of a weighted graph have distinct weights, the graph has only one minimum spanning tree.

Earlier in this section, we discussed the problem of networking the computers in a building, where the cable must be installed alongside the already existing wiring in the building. The eight computers and current wiring were laid out as shown in Figure 3.61. This situation can be easily represented by a weighted graph, where the vertices represent the computers and the edges represent the possible places that the cable can be installed. If we use Prim's algorithm to find a minimum spanning tree, one of the possible resulting minimum spanning trees corresponds to the network shown in Figure 3.63. (You should check this yourself.) Therefore, the minimum amount of cable that can be used to network the computers is the weight, 62 yards, of this spanning tree.

In the next example, we see another application of networking.

EXAMPLE 2 *Bike Trails*

Suppose five towns that are connected by roads would like to build bike trails on some of the roads in such a way that a biker could travel from any of the five towns to another along bike trails, though possibly by an indirect route. If the towns and the current roads and their lengths in miles are as shown in Figure 3.73, along which roads should the bike trails be built in order to keep the total length of the trails to a minimum?

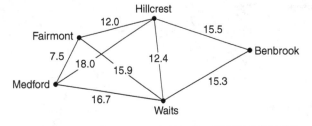

FIGURE 3.73

The towns can be considered vertices and the roads edges of a weighted graph. To keep the total length of the trails to a minimum, they should be built along the roads forming a minimum spanning tree in this graph. If we use Prim's algorithm starting at the town of Fairmont, then we first connect Fairmont and Medford. Next, we connect Hillcrest to Fairmont, then Waits to Hillcrest. Last of all, we connect Benbrook to Waits. The resulting minimum spanning tree is as shown in Figure 3.74.

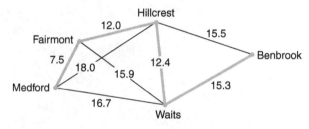

FIGURE 3.74

E X E R C I S E S FOR SECTION 3.3

In Exercises 1–6, determine whether or not the subgraph colored blue is a tree. If it is not a tree, explain why not. If it is a tree, determine whether or not it is a spanning tree.

1.
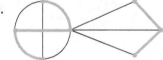

2.

3.

4.

5.

6.

In Exercises 7 and 8, find three different spanning trees of the graph.

7.

8.

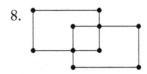

In Exercises 9 and 10, find all possible spanning trees of the graph.

9.

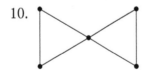

10.

In Exercises 11–18, use Prim's algorithm to find a minimum spanning tree for the given weighted graph and find the weight of this tree.

11.

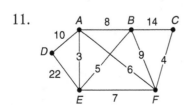

12.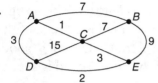

13.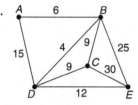

14.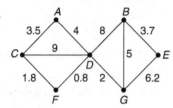

15.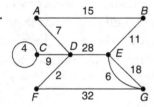

16.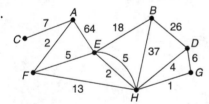

17.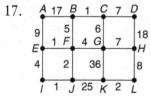

18.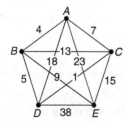

19. Six small towns are currently connected by dirt roads as illustrated in the figure, where the numbers indicate the lengths of the roads in miles. Which of these roads should be paved so it would be possible to travel from any town to another along only paved roads and so the total length of the roads paved is kept to a minimum? What is the minimum length?

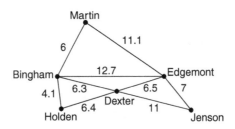

20. Seven buildings on a high school campus are connected by the sidewalks shown in the figure, where the numbers indicate the length of each sidewalk in yards. A heavy snow has fallen, and the groundskeeper must shovel the sidewalks quickly. The groundskeeper therefore decides to shovel as little as possible and still ensure that a person walking from any of the buildings to another will be able to do so along only cleared sidewalks and through buildings. Draw a graph corresponding to this situation, and indicate those sidewalks the worker should shovel. Also, give the total length of the sidewalks the groundskeeper should shovel.

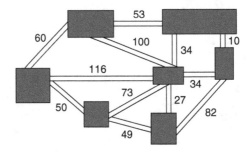

21. Suppose a small airline in the northwestern United States currently has flights between some of the cities in the figure, where the mileage between the cities with flights between them is indicated on the edge between the cities. The

airline wants to reduce the total number of different flights it offers but wants to be able to provide travel between any two cities, where in some cases it may be necessary to take more than one flight. Which of the current flights should be retained in order to do this in such a way that the total length of the remaining flights is as short as possible?

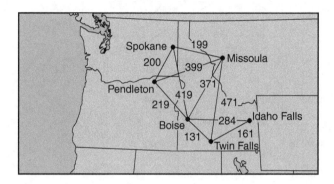

22. An irrigation system is to be installed. The main water source is located at vertex A in the graph and the other vertices represent the proposed sprinkler head locations. The edges indicate all possible choices for installing pipes. For the irrigation system to work, each sprinkler head must be connected to the main water source through one or more pipes. The costs of installing the pipes, in dollars, are given as the weights of the corresponding edges. Find the least expensive way of installing the irrigation system, and give the cost.

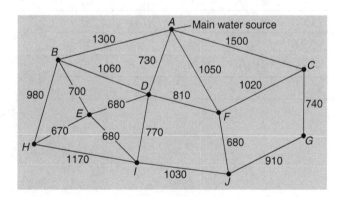

23. The highway distances, in miles, between the cities of Boston, Cincinnati, Cleveland, Detroit, New York, and Philadelphia are given as the weights of the edges between each pair of cities in the graph. If a high-speed fiber optic cable is to be installed along these highways, where should the cable be installed so the

minimal amount of cable is used to connect the cities? What is the total length of cable needed?

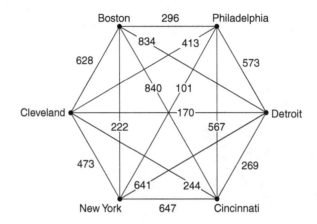

24. Five computers in an office, which we will label *A*, *B*, *C*, *D*, and *E*, must be networked with cable. The cost of connecting each pair of computers by cable is given in the table. Draw a graph corresponding to this situation. Between which computers should the cable be installed so as to keep the total cost of networking to a minimum? What is the total cost of this least expensive network?

	A	B	C	D	E
A	—	$120	$250	$180	$225
B	$120	—	$85	$130	$85
C	$250	$85	—	$200	$175
D	$180	$130	$200	—	$220
E	$225	$85	$175	$220	—

25. A company needs to provide a computer network between its four main offices in Baltimore, Charlotte, Chicago, and Los Angeles. Suppose the monthly cost for leasing a computer line between the four cities is as shown in the table. Draw a graph corresponding to this situation, and find the minimum cost network. What is the total monthly cost of this network?

	Baltimore	Charlotte	Chicago	Los Angeles
Baltimore	—	$820	$3500	$2000
Charlotte	$820	—	$1100	$1750
Chicago	$3500	$1100	—	$1300
Los Angeles	$2000	$1750	$1300	—

WRITING EXERCISES

1. Explain why the traveling salesman problem is an inappropriate model for a round-trip tour of all the planets in the solar system.

2. Explain why, in building some networks, we might want more connections than just those of a minimal spanning tree.

3. Give an example of a weighted graph modeling a real situation where for any two vertices in the graph connected by an edge, the weight of the edge between the two vertices is at most the sum of the weights of the edges of any other path connecting the two vertices. Would you expect this to be true for all weighted graphs modeling real situations? Explain why or why not.

4. In Exercise 19 of Section 3.3, we considered the problem of paving the roads between six towns that are currently connected by dirt roads in such a way that it would be possible to travel from any town to another along only paved roads and so the total length of the roads paved is kept to a minimum. Suppose you are to present the proposed solution of paving the minimum spanning tree, shown in the figure, to the assembled councils of the towns. What sorts of comments or questions would you anticipate, and how would you respond to them? How might you propose that the cost of paving be allocated, and how would you justify your proposal?

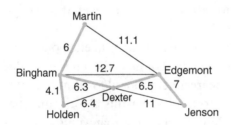

5. The version of the traveling salesman problem we have considered is called the *symmetric traveling salesman problem* because the weight of the path from vertex A to vertex B is the same as that from B to A.

   (a) Can you think of some applications for which the weights would not be equal?

   (b) Can the nearest neighbor algorithm and the greedy algorithm be applied in this asymmetric case? If so, what adaptations must be made? If not, why not?

PROJECTS

1. Invent an approximate algorithm for solving the traveling salesman problem that is different from the nearest neighbor and greedy algorithms. Test your algorithm on several graphs and compare it to the nearest neighbor and greedy algorithms.

2. The basic idea of an algorithm may be applied in several different settings. For instance, the nearest neighbor algorithm and Prim's algorithm are essentially the same algorithm applied in the settings of the traveling salesman problem and of finding the minimum spanning tree.

   (a) Describe how to adapt the greedy algorithm from the context of the traveling salesman problem to the context of finding the minimum spanning tree. (This algorithm should be different from Prim's algorithm.)

   (b) By trying a variety of examples, see if you think your algorithm always gives the minimum spanning tree.

3. Many calculators have a random number generator. Use one that does, or the random number generator on a computer, to generate random weights for each of the six edges of a complete graph on vertices A, B, C, and D. (Often values will be between 0 and 1, but any range of positive numbers will do.) For each of 20 trials (more if you are part of a group), compute the weight of the minimal circuit, the weight of the circuit obtained by the nearest neighbor algorithm starting at vertex A, and the weight of the circuit obtained by the greedy algorithm. Find the average of each of these three values over the 20 trials. What percentage of the time does each algorithm give a circuit of minimal total weight? What does this suggest to you about choosing between these methods for solving problems arising in practice?

4. (a) What is the least number of edges needed on a graph with n vertices to be sure that all vertices are connected even if one edge is destroyed (say, by a natural disaster)?

   (b) What is the least number of edges needed on a graph with n vertices to be sure that all vertices are connected even if two edges are destroyed?

5. Consider either the nearest neighbor algorithm or Prim's algorithm. In writing a computer program to implement this algorithm, what quantities need to be stored in memory? Describe the steps that would have to be programmed. Write a program implementing your choice of algorithm.

KEY TERMS

CHAPTER 3 / REVIEW TEST

In Exercises 1–4, determine whether or not the graph has an Eulerian circuit, and if it does, give one. If it does not, explain why not. If the graph does not have an Eulerian circuit, determine if it has an Eulerian path, and if it does, give one. If it does not, explain why not.

1.

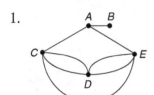

2.

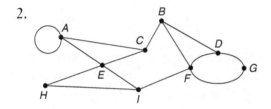

3.

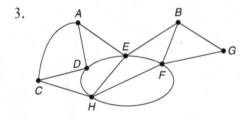

4.

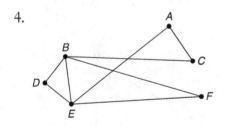

In Exercises 5 and 6, find an eulerization of the graph.

5.

6.

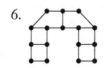

In Exercises 7 and 8, find a Hamiltonian circuit, if possible, in the given graph. If it is not possible to find a Hamiltonian circuit, explain why not.

7.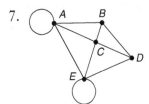

8.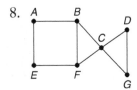

In Exercises 9 and 10, find an approximate solution to the traveling salesman problem for a circuit starting at vertex A on the given weighted graph and find the weight of the circuit using (a) the nearest neighbor algorithm, (b) the greedy algorithm.

9(a).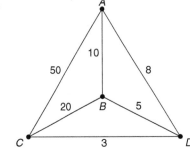

(b).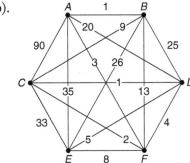

10(a).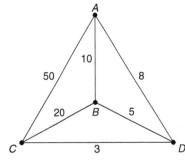

(b).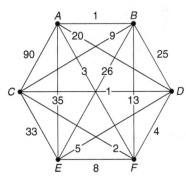

In Exercises 11 and 12, use Prim's algorithm to find a minimum spanning tree for the given weighted graph, and find the weight of this tree.

11.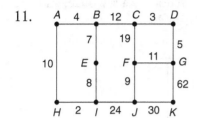

12.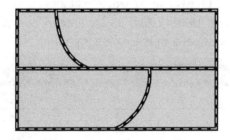

13. A political campaign volunteer wants to deliver flyers to each of the homes in the neighborhood pictured. The streets on the outside of the neighborhood have houses on only one side, and the interior streets have houses on both sides. The volunteer plans to park her car at one of the street intersections and walk down each of the interior streets twice (once for each side) and each of the outside streets once. Draw a graph corresponding to this situation. Determine whether or not the graph has an Eulerian circuit, and if so, give one. If it does not, explain why not.

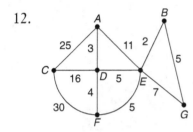

14. A burglar, when apprehended, claimed that he entered a room in the house shown in the figure by forcing a window open at the location marked **w.** He then went through the house, passing through each door exactly once, and left through the back door at **x.** Why must the burglar be lying?

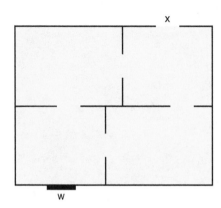

15. A trucker starting in Denver must travel to Cheyenne, Chicago, Des Moines, and Minneapolis, and then return to Denver in one trip. The road mileage between these cities is listed in the following table:

	Cheyenne	Chicago	Denver	Des Moines	Minneapolis
Cheyenne	—	954	100	627	870
Chicago	954	—	996	327	405
Denver	100	996	—	669	920
Des Moines	627	327	669	—	252
Minneapolis	870	405	920	252	—

(a) Draw a weighted graph corresponding to this situation.

(b)   Use the nearest neighbor algorithm to find a travel route, and find the length of this route.

(c)   Use the greedy algorithm to find a travel route, and find the length of this route.

16.   The homes in a rural community, indicated by vertices, and the possible choices for installing a gas pipeline system with the corresponding costs, in dollars, are shown in the figure. Find the cheapest way of installing the pipeline so that all of the homes are connected to it, and give the cost.

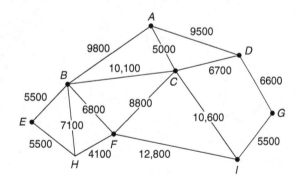

17. For what values of n does a complete graph with n vertices have an Eulerian circuit?

SUGGESTED READINGS

Bern, Marshall W., and Ronald L. Graham. "The Shortest-Network Problem." *Scientific American* 260:1 (January 1989), 84–89. Introduction to the problem of finding minimal networks that allow the inclusion of Steiner points.

Euler, Leonhard. "The Koenigsberg Bridges," trans. James Newman. *Scientific American* 189:1 (July 1953), 66–70. English translation of Euler's original paper on the Königsberg bridge problem.

Lewis, Harry R., and Christos Papadimitriou. "The Efficiency of Algorithms." *Scientific American* 238:1 (January 1978), 96–109. Discussion of efficient algorithms in the context of graph theory.

Rosen, Kenneth H. *Discrete Mathematics and Its Applications, 7th Edition.* New York: McGraw-Hill, 2012. One of many discrete mathematics textbooks that include a somewhat more technical, but still accessible, treatment of graph theory.

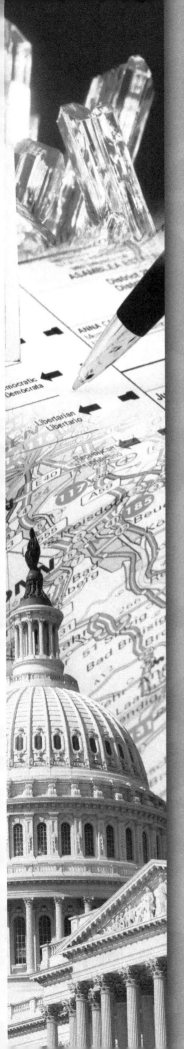

Compound eye of a wasp.

Geometrical patterns and symmetry can be seen in many objects in nature, for instance in the compound eyes of a fly and in a group of soap bubbles. Because of the beauty and usefulness of such patterns, we also find them in man-made objects. Consider, for example, the patterns that appear on the Dome of the Rock in Jerusalem, a beautiful masterpiece of Islamic art and architecture built in 692 A.D.

In this chapter we will see the geometry that lies behind such patterns and shapes. We begin with a review of polygons and angle measure. We then apply these ideas to explore some of the kinds of patterns that can arise in tiling a flat surface with polygons. We conclude the chapter by looking at regular and semiregular polyhedra. Throughout the chapter we will see that the ways that shapes can be fitted together are governed by simple geometric relations.

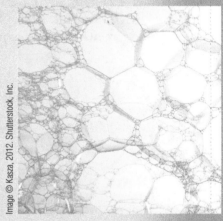

Soap bubbles.

The Dome of the Rock in Jerusalem (known in Arabic as the Qubbat as-Sakhra).

TILINGS
AND POLYHEDRA

4.1 POLYGONS

Many patterns in nature and art are composed of one or more geometrical shapes fitted together in a pattern. The Roman mosaic in Figure 4.1 is an example of such a pattern. In this tiling, as in many patterns, the basic geometric shapes are polygons.

FIGURE 4.1 A Roman mosaic.

We use the word **polygon** to refer to a *simple* polygon, a connected figure in the plane consisting of a finite number of line segments each of whose endpoints intersects an endpoint of exactly one of the other line segments and such that there are no other points where two of the line segments intersect. The shapes in Figures 4.2(a) and 4.2(b) are both polygons.

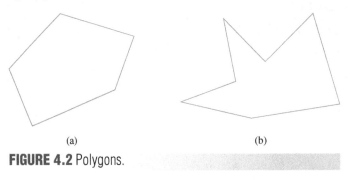

(a)   (b)

FIGURE 4.2 Polygons.

The shape in Figure 4.3(a) is not a polygon because two of the line segments have endpoints that do not intersect the endpoint of another line segment, and the shape in Figure 4.3(b) is not a polygon by our definition because it has sides that intersect in a place other than their endpoints. We will also use the word polygon to refer to a polygon along with its interior because we will have no need to distinguish between whether or not the interior is included.

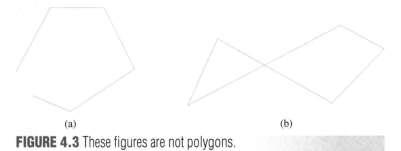

(a)   (b)

FIGURE 4.3 These figures are not polygons.

Each of the line segments forming a polygon is called a **side**, and each point at which two sides touch is called a **vertex**. (The plural of vertex is vertices.) The name of a polygon is based on how many sides or angles it has. The names for several different polygons are given in Table 4.1.

TABLE 4.1 The Names of Polygons

Number of Sides	Name of the Polygon	Example
3	Triangle	
4	Quadrilateral	
5	Pentagon	
6	Hexagon	
7	Heptagon	
8	Octagon	
9	Nonagon	
10	Decagon	
12	Dodecagon	

A polygon with *n* sides, especially when it is not one of the ones included in Table 4.1, is called an ***n*-gon**. For instance, a polygon with 17 sides is called a 17-gon.

Polygons can be classified as convex or concave. A polygon is called **convex** if every line segment drawn between two points in the polygon lies completely inside the polygon. A polygon that is not convex is called **concave**. The polygon in Figure 4.4 is convex. To convince yourself of this, try to connect any two points within the polygon, and you will see that the connecting line segment stays inside the polygon.

When we try to connect the two points *A* and *B* inside the polygon in Figure 4.5, the line segment connecting them falls outside the polygon. Therefore, the polygon is concave.

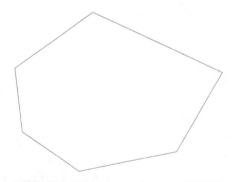

FIGURE 4.4 A convex polygon.

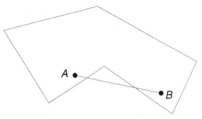

FIGURE 4.5 A concave polygon.

FIGURE 4.6 The interior angles of a polygon.

The inside angle at which two sides of a polygon meet is called an *interior angle*. In Figure 4.6, the interior angles of the polygon are marked. For brevity, we call an interior angle simply an **angle**, and it is to be understood that we are referring to an interior angle. The angles of a convex polygon all measure at most 180°, whereas a concave polygon has at least one angle measuring more than 180°.

We are particularly interested in looking at the most symmetric of polygons, called the regular polygons. They are defined as follows:

A **regular polygon** is a polygon with sides of equal length and angles of equal measure.

The first six regular polygons are shown in Figure 4.7.

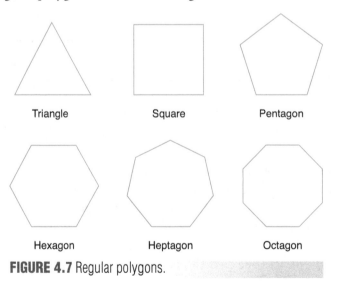

| Triangle | Square | Pentagon |

| Hexagon | Heptagon | Octagon |

FIGURE 4.7 Regular polygons.

A regular triangle is more commonly called an equilateral triangle, and a regular quadrilateral is a square. Notice that a regular *n*-gon has rotational symmetry in the sense that if we spin it one *n*th of a full turn about its center, it appears exactly as it did before the turn. It is important to note that for a polygon to be regular, it is not

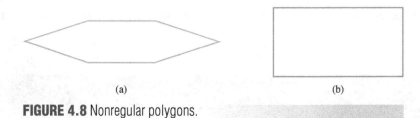

(a)   (b)

FIGURE 4.8 Nonregular polygons.

enough to require that the edges be of equal length. For instance, the hexagon in Figure 4.8(a) is not regular for, although the sides are of equal length, the angles are not of equal measure. Notice that this hexagon does not have the kind of rotational symmetry that a regular hexagon has. It is also not enough to require that angles be of equal measure. For instance, the rectangle in Figure 4.8(b) is a quadrilateral with angles of equal measure, but it is not regular. It is not the most symmetrical of quadrilaterals.

In the next section, we will construct patterns involving regular polygons, and we will need to know the measure of an angle of a regular polygon. To find a formula for this, we first use the fact that the sum of the angles of any triangle is 180° to find a formula for the sum of the angles of any *n*-gon, not necessarily regular. Let's first look at a convex quadrilateral. If we draw the line segment from any vertex to its opposite vertex, then the quadrilateral is divided into two triangles as shown in Figure 4.9(a). From this picture, we see that the sum of the angles of the quadrilateral is equal to the sum of the angles of the two triangles. Thus, it follows that the sum of the angles of a quadrilateral is 2 · 180° = 360°. Similarly, for any convex pentagon, we can draw two straight lines from any vertex to the vertices opposite it as in Figure 4.9(b), and this will divide the pentagon into three triangles. The sum of the angles of these three triangles is equal to the sum of the angles of the pentagon, so we see that the sum of the angles of a pentagon is given by 3 · 180° = 540°. The same idea works for any convex hexagon, and because in this case the hexagon can be divided into four triangles as in Figure 4.9(c), we find that the sum of the angles of a hexagon is 4 · 180° = 720°.

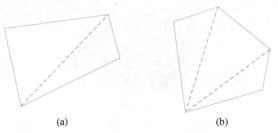

(a)   (b)   (c)

FIGURE 4.9 Breaking convex polygons into triangles.

FIGURE 4.10 Breaking a concave hexagon into triangles.

In a similar way, any convex *n*-gon can be broken into *n* − 2 triangles. By looking at a few cases, you should be able to convince yourself that it is also possible to break any concave *n*-gon into *n* − 2 triangles. For instance, the concave hexagon in Figure 4.10 has been broken into 4 triangles. Therefore, to find the sum of the angles of any n-gon, a similar argument can be used to arrive at the following result:

The sum of the angles of any n-gon is $(n-2)\,180°$.

EXAMPLE 1 *Sum of the Angles of an n-gon*

Find the sum of the angles of a dodecagon.

SOLUTION:

A dodecagon has 12 sides, so using the formula for the sum of the angles of a 12-gon, we see that the angle sum is given by

$$(12 - 2)180° = 10 \cdot 180° = 1800°. \qquad \blacksquare$$

Because the n angles of a regular n-gon are all equal, we can find the measure of each angle of a regular n-gon by dividing the sum of the angles by n to get

$$\frac{(n-2)180°}{n} = \frac{180°n - 360°}{n} = 180° - \frac{360°}{n}.$$

Therefore, we have the following formula:

Angle of a Regular *n*-gon
The measure of an angle of a regular n-gon is
$$180° - \frac{360°}{n}.$$

EXAMPLE 2 *Measure of the Angle of a Regular n-gon*

Find the measure of the angle of a regular

(a) pentagon,
(b) 14-gon.

SOLUTION:

(a) Using the formula for the measure of the angle of a regular *n*-gon with *n* = 5, we see that the measure of the angle of a regular pentagon is

$$180° - \frac{360°}{5} = 180° - 72° = 108°.$$

(b) The measure of the angle of a regular 14-gon is

$$180° - \frac{360°}{14} \approx 180° - 25.7143° = 154.2857°.$$ ■

E X E R C I S E S FOR SECTION 4.1

In Exercises 1–8, determine whether the figure is a polygon. If it is not, explain why not.

1.

2.

3.

4.

5.

6.

7.

8.

In Exercises 9–16, determine whether the given polygon is convex or concave, whether or not it appears regular, and state the polygon's name.

9.

10.

11.

12.

13.

14.

15.

16.

In Exercises 17–22, find the sum of the angles of the given polygon.

17. decagon   18. octagon
19. 17-gon   20. heptagon
21. nonagon   22. 16-gon

In Exercises 23–28, find the measure of an angle of the given polygon.

23. regular octagon   24. regular hexagon
25. regular heptagon   26. regular 11-gon
27. regular 30-gon   28. regular dodecagon

29. Three of the four angles of the following polygon are given. Find the measure of the remaining angle.

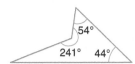

30. Four of the five angles of the following polygon are given. Find the measure of the remaining angle.

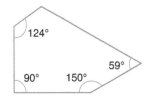

31. Three of the angles of a pentagon are 35°, 42°, and 93°. The other two angles are equal. What is the measure of each of the other angles?

32. Two of the angles of a quadrilateral are 37° and 161°. One of the two remaining angles is twice as large as the other remaining angle. What are the measures of the two remaining angles?

4.2 | TILINGS

Certain sorts of polygons and other basic shapes can be placed side-by-side to form geometric patterns that completely cover a flat surface. These kinds of patterns can be found in the tiling of floors, in fabric and wallpaper designs, in nature, and in art. The design of the quilt we see in Figure 4.11 is based on a pattern of regular hexagons, whereas in Figure 4.12 we see a floor tiling pattern made up of squares and hexagons that are not regular. The patterns in both of these pictures can be extended in every direction to cover as large an area as we wish.

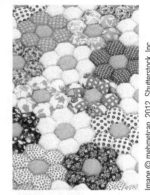

FIGURE 4.11

FIGURE 4.12

We think of the basic shapes that are used in forming these patterns as tiles. A pattern in which one or more repeated tiles are used to cover a plane without leaving gaps or overlapping and that can be extended forever in every direction is called a **tiling** or **tessellation**.

Tilings that use only one size and shape of tile, where the tile may be used face up or face down, are called **monohedral** tilings. We will look at both monohedral tilings and tilings using more than one kind of tile. However, we will restrict our attention to tilings for which each edge of a tile coincides exactly with the edge of a bordering tile. Tilings of this kind are called **edge-to-edge**. In Figure 4.13(a), we see a tiling that is edge-to-edge, and in Figure 4.13(b) we see one that is not edge-to-edge.

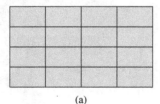

(a)

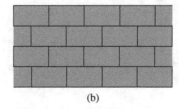

(b)

FIGURE 4.13 (a) Edge-to-edge   (b) Not edge-to-edge

The **vertices** of an edge-to-edge tiling are the vertices of the tiles, each of which belongs to more than one tile. In Figure 4.14, we see an edge-to-edge tiling in which each vertex belongs to three, four, or five tiles.

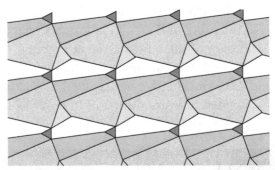

FIGURE 4.14 An edge-to-edge tiling by triangles, quadrilaterals, and pentagons.

Regular Tilings

We begin by asking which regular polygons can be used to tile the plane by themselves. You have almost certainly seen tilings formed using equilateral triangles, squares, or regular hexagons, as shown in Figure 4.15.

FIGURE 4.15 Tilings by equilateral triangles, squares, and regular hexagons.

Notice that in each case the tiles fit together exactly and the pattern can be extended in every direction. These three tilings are regular tilings, which are defined as follows:

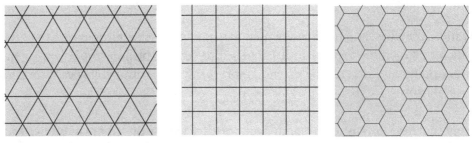

A **regular tiling** is an edge-to-edge monohedral tiling for which the tiles are regular polygons.

FIGURE 4.16

Are the three regular tilings in Figure 4.15 the only possible regular tilings? To answer this question, it is helpful to understand why these three regular tilings work. Let's first look at the tiling with equilateral triangles. Because the triangles must fit about each vertex without overlapping or leaving any gaps as shown in Figure 4.16, we know that the angle of an equilateral triangle must divide 360°, the total angle about a vertex, evenly. In fact, the angle of an equilateral triangle is 60° and 360°/60° = 6, the number of triangles about each vertex.

Similarly, the angle of a square is 90°, and 90° divides 360° evenly, giving 360°/90° = 4, the number of squares about each vertex as seen in Figure 4.17.

FIGURE 4.17

For the case of the regular hexagons, we first use the formula for the angle of a regular *n*-gon to find that the angle of a regular hexagon is

$$180° - \frac{360°}{6} = 180° - 60° = 120°.$$

FIGURE 4.18

The angle 120° divides 360° evenly, giving 360°/120° = 3, which is the number of hexagons about each vertex as we see in Figure 4.18.

If any other *n*-gon could be used to make a regular tiling, then it must be true that the angle of this *n*-gon divides 360° evenly. The angle of a regular pentagon is given by

$$180° - \frac{360°}{5} = 180° - 72° = 108°,$$

which does not divide 360° evenly. The angle of any regular polygon with more than six sides is larger than 120°, the angle of a regular hexagon, but smaller than 180°. Because 360°/180° = 2 and 360°/120° = 3, we see that when 360° is divided by any number between 180° and 120°, the result is a number between 2 and 3, and thus not an integer. Therefore, we have the following fact:

> The only regular tilings are those with equilateral triangles, with squares, or with regular hexagons.

Semiregular Tilings

Having found all of the regular tilings, we will now look for edge-to-edge tilings that involve more than one kind of regular polygon. As with regular tilings, the sum of the angles of the polygons around each vertex must add up to exactly 360°. The arrangement of polygons around a vertex determines the vertex type.

> Two vertices are said to be of the same **vertex type** if the same number and kind of polygons are arranged about them in the same order, except that one arrangement may be clockwise whereas the other may be counterclockwise.

branching OUT

THE HEXAGON PATTERN OF HONEYBEE COMBS

triangular array with the cylinders just touching as shown in the figure.

One of the places in nature where the hexagon regular tiling pattern appears is in the familiar honeybee comb. It seems quite remarkable that the bees construct such a symmetric and beautiful pattern. However, the pattern emerges very simply as a result of the manner in which the bees build the comb. A comb is constructed by a group of bees, each one building a cylinder of wax about itself. The cylinders are about the same size and the bees arrange themselves in a

This is an efficient use of space in packing the cylinders. That the bees arrange themselves in this pattern is not the result of careful planning but instead follows as a result of the bees exerting approximately equal pressures in trying to make their own circles as large as possible. Because the wax is soft, the cylinders then flow together to form the regular hexagonal pattern.

In Figure 4.19(a) and (b), the indicated vertices are not of the same vertex type because, even though the number and types of polygons are the same, the arrangements of the polygons about the vertices are different.

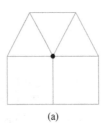
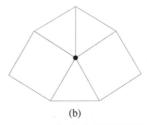

(a)   (b)

FIGURE 4.19 Vertices of different types.

The indicated vertices in Figure 4.20(a), (b), and (c) are all of the same vertex type, even though the clockwise order of the polygons about the vertices in Figure 4.20(a) and (b) is the counterclockwise order in Figure 4.20(c).

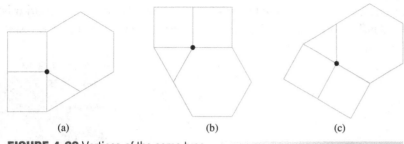

(a)   (b)   (c)

FIGURE 4.20 Vertices of the same type.

We now have the following definition:

A **semiregular tiling** is an edge-to-edge tiling by at least two different regular polygons such that all of the vertices are of the same type.

There are only eight semiregular tilings. As with regular tilings, these can be found first by looking for sets of regular polygons that will fit around a single vertex and then by checking whether these patches of tiles can be extended to cover the plane. We will find one of these semiregular tilings in the following example and will look for others in the exercises.

EXAMPLE 1 *Finding a Semiregular Tiling*

Find a semiregular tiling that has three tiles about each vertex, at least one of which is an octagon.

SOLUTION:

The angles of the three regular polygons about each vertex must add up to 360°, and the angle of a regular octagon is 180° − 360°/8 = 135°. Therefore, the angles of the remaining two polygons about each vertex must add up to 360° − 135° = 225°. Because 225°/2 = 112.5°, one of the two remaining tiles about each vertex has to have an angle of at most 112.5°. Recall that a regular hexagon has angles measuring 120°, and regular n-gons with $n > 6$ have even larger angles. Therefore, we can conclude that one of the two remaining polygons must be a regular pentagon with 108° angles, a square with 90° angles, or an equilateral triangle with 60° angles. If one of the polygons is a regular pentagon, the third polygon must have angles measuring 225° − 108° = 117°. But there is no regular polygon with an angle of this size. If one of the polygons is a square, the third polygon must have angles measuring 225° − 90° = 135°, the measure of the angle of a regular octagon. There is only one way to arrange two octagons and a square about a vertex and it is shown in Figure 4.21. By forming a tiling where every vertex is of this type, we get the familiar tiling in Figure 4.22.

FIGURE 4.21

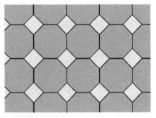

FIGURE 4.22

TECHNOLOGY *tip*

When trying to construct tilings such as in Example 1, it is helpful to use a computer drawing program to construct the tiles. These tiles can then be easily duplicated and moved around. If a drawing program is not available to you, the same idea can be carried out with paper cutouts of tiles.

FIGURE 4.23

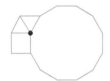

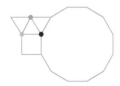

FIGURE 4.24

It is important to note here that even when a set of regular polygons fits about a vertex, we cannot be sure that they will give a semiregular tiling. For instance, two equilateral triangles, one square, and one regular 12-gon fit about a vertex as illustrated in Figure 4.23. However, a semiregular tiling with vertices of this type is not possible. To see why, consider what happens when we try to continue this tiling so that all the vertices are of the same type. In Figure 4.24, we see that another triangle must be placed at the blue vertex so that the blue vertex is of the same type as the original black vertex. However, this forces the green vertex to be of a different type because it includes more than two triangles.

There are 21 different vertex types consisting of regular polygons arranged about a vertex. However, only 11 of these give either regular or semiregular tilings. We will explore these further in the exercises. Note that the order in which the polygons are arranged about a vertex may be important. For instance, you are asked to show in an exercise that it is not possible to make a semiregular tiling where the vertices are of the type shown in Figure 4.20, but a different arrangement of the same polygons about a vertex will give a semiregular tiling.

Although we have focused our discussion of tilings by regular polygons on the regular and semiregular tilings, these 11 tilings are certainly not the only possibilities for edge-to-edge tilings of the plane with regular polygons. When we take away the restriction of having each vertex be of the same type, the possibilities are limitless. Figure 4.25, which is a tiling with regular polygons where three different vertex types appear, is just one of the infinite number of possibilities.

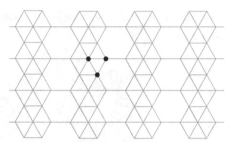

FIGURE 4.25 An edge-to-edge tiling by regular polygons that is not a regular or semiregular tiling. The three different vertex types are marked by the dots.

Tilings with Nonregular Polygons

We know that the only regular polygons that can be used to form an edge-to-edge monohedral tiling are equilateral triangles, squares, and regular hexagons. We now look at the more general idea of monohedral tilings where the tile used is in the shape of a polygon that is not regular.

It is possible to form an edge-to-edge tiling of the plane with any kind of triangle. One way to do this is to attach two copies of the triangle together to form a parallelogram—that is, a quadrilateral with opposite sides parallel to one another, as shown in Figure 4.26(a). These parallelograms can then be used to tile the plane as shown in Figure 4.26(b).

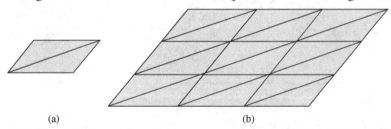

(a)   (b)

FIGURE 4.26 Tiling with a triangle.

Next we might ask whether it is possible to form an edge-to-edge tiling of the plane with any kind of quadrilateral. As with triangles, the answer is yes. A tiling with any quadrilateral can be formed by first attaching two copies of the quadrilateral in such a way that a hexagon with opposite sides parallel is formed as shown in Figure 4.27(a). These hexagons can then be used to form an edge-to-edge tiling as shown in Figure 4.27(b).

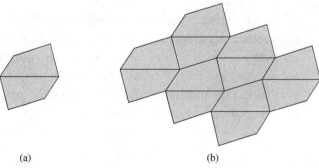

(a)   (b)

FIGURE 4.27 Tiling with a convex quadrilateral.

This method of tiling with a quadrilateral also works for concave quadrilaterals, as illustrated in Figures 4.28(a) and 4.28(b).

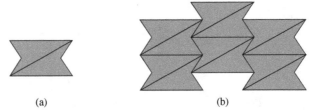

(a)   (b)

FIGURE 4.28 Tiling with a concave quadrilateral.

Because it is possible to tile with regular hexagons, it is natural to wonder whether we can tile with any hexagon. It turns out that some but not all hexagons can be used to tile the plane. In 1918, mathematician Karl Reinhardt showed that only three different types or families of convex hexagons can tile the plane. They are shown in Figure 4.29. Notice, however, that tiling with a hexagon of Type 2 requires that some of the tiles be flipped over so that they are laying face down. Recall that this is allowed in the definition of a monohedral tiling.

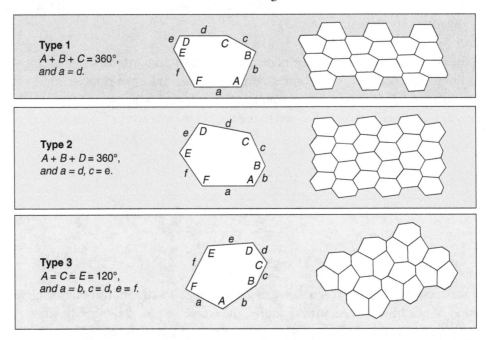

Type 1
$A + B + C = 360°$, and $a = d$.

Type 2
$A + B + D = 360°$, and $a = d$, $c = e$.

Type 3
$A = C = E = 120°$, and $a = b$, $c = d$, $e = f$.

FIGURE 4.29 The three families of convex hexagons that can tile the plane.

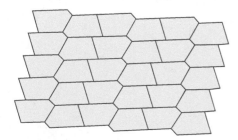

FIGURE 4.30 A tiling by a pentagon with two sides parallel.

We know that the regular pentagon cannot tile the plane, but are there other pentagons that can? The situation is similar to that for hexagons in that some but not all pentagons can be used in a monohedral tiling. For instance, any pentagon with two sides parallel to one another can be used to tile the plane. An example of such a tiling is shown in Figure 4.30.

For convex pentagons, it is now known that at least 14 different families can tile the plane, but not all of these families give rise to edge-to-edge tilings. It is not known whether there are any others. Interestingly, four of the families of pentagons that give tilings were discovered in the 1970s by amateur mathematician Marjorie Rice, who had no formal mathematics education beyond high school. One of the tilings she discovered is shown in Figure 4.31.

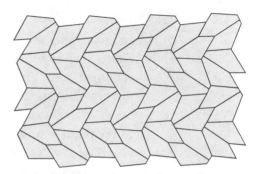

FIGURE 4.31 A tiling by pentagons discovered by Marjorie Rice in 1976. (This tiling is not edge-to-edge.)

What about tiling with n-gons, where $n \geq 7$? It is possible to tile the plane with some concave n-gons with seven or more sides. A beautiful spiral monohedral tiling by concave 9-gons, discovered by Heinz Voderberg in 1936, is shown in Figure 4.32. However, in 1927, Karl Reinhardt proved that it is not possible to form a monohedral tiling of the plane with a convex n-gon for $n \geq 7$.

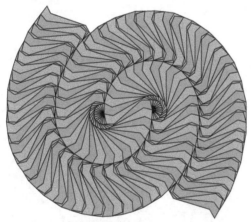

FIGURE 4.32 A monohedral tiling by concave 9-gons discovered by Heinz Voderberg.

Tilings with Other Shapes

Some of the most beautiful tilings involve tiles that are not in the shape of polygons. Dutch artist M. C. Escher (1898–1972) formed tilings with animals and other creatures. Figures 4.33 and 4.34 are tilings similar to Escher's.

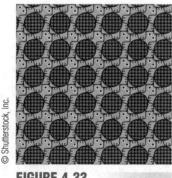

FIGURE 4.33

FIGURE 4.34

Escher designed the tiles used in his work by starting with a simple polygon or polygons and modifying them in a systematic way. Escher was not the first to use recognizable figures such as animals and leaves in tilings. Examples from the end of the nineteenth century precede his work, but it was Escher who popularized the style. We will look at just two of the methods for modifying polygons to form Escher-like monohedral tilings.

In the first method, we start with a rectangle as in Figure 4.35(a). One side—in our case the top side—of the rectangle is modified as in Figure 4.35(b), and this modification is copied on the opposite side, as illustrated in Figure 4.35(c). Thus the tiles fit into one another like a jigsaw puzzle. We can then modify one of the two remaining sides (or leave them unmodified) and copy that change on the opposite side as illustrated in Figure 4.35(d). The resulting tile can now be used to tile the plane as shown in Figure 4.35(e).

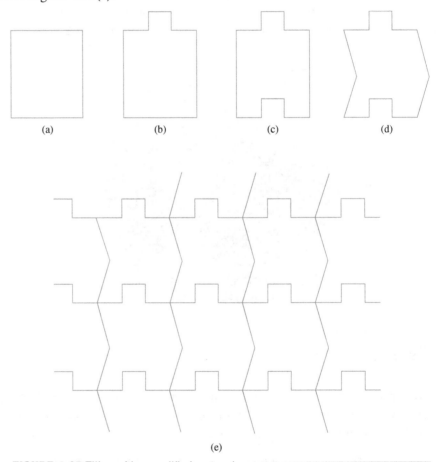

FIGURE 4.35 Tiling with a modified rectangle.

With a little bit of embellishment of the tile, we get Figures 4.36(a) and 4.36(b).

Another method for forming a tiling is to start with any triangle as shown in Figure 4.37(a). We first modify one half of one side of the triangle, in our case the upper right side as in Figure 4.37(b), and then carry this modification over to the other half of the side by rotating the side about its center 180° as shown in Figure 4.37(c). A similar modification can be made to each of the other two sides, if desired, as shown in Figures 4.37(d) and 4.37(e). A tiling can now be made from the resulting tile as shown in Figure 4.37(f).

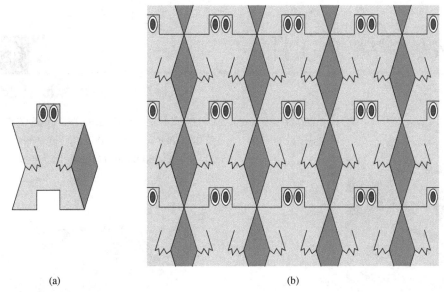

(a)   (b)

FIGURE 4.36 Embellished tiling by a modified rectangle.

Notice that this tiling is based on the kind of tiling by a triangle shown in Figure 4.26. Again, we add a creative touch as in Figure 4.38(a) and the tiles can now be fitted together to arrive at the tiling in Figure 4.38(b).

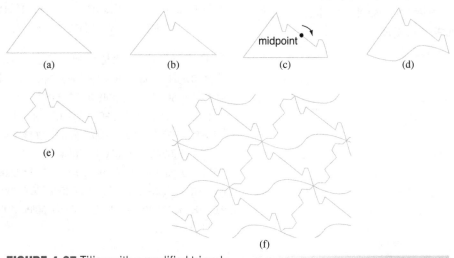

(a)   (b)   (c)   (d)

(e)

(f)

FIGURE 4.37 Tiling with a modified triangle.

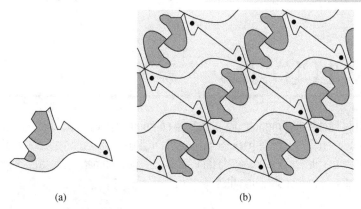

(a)   (b)

FIGURE 4.38 Embellished tiling by a modified triangle.

branching OUT

NAPOLEON AND TILINGS

Napoleon Bonaparte (1769–1821).

Napoleon's Theorem can be proved by using the fact that the plane can be tiled using the given triangle and the attached equilateral triangles as shown in figure (b).

(b)

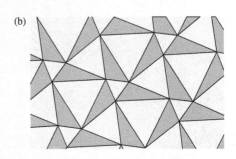

Napoleon Bonaparte, general and emperor of France, was quite interested in geometry. A result in geometry—attributed to him and known as Napoleon's Theorem—says that given any triangle, if we attach equilateral triangles along each side, then the centers of these attached triangles form an equilateral triangle, as shown in figure (a).

(a)

The theorem is proved by showing that the centers of the equilateral triangles in this tiling form the vertices of a regular tiling with triangles.

There is some doubt as to whether Napoleon actually proved the theorem named after him. According to a story, sometime before he became the emperor of France, Napoleon had a discussion with the French mathematicians Lagrange and Laplace. In the course of the conversation, Laplace told him "The last thing we want from you, general, is a lesson in geometry." Apparently Napoleon took the comment well, because Laplace later became Napoleon's chief military engineer.

TECHNOLOGY *tip*

When constructing Escher-like monohedral tilings, a computer drawing program is quite helpful. There are also commercial software packages available that partially automate some of the steps needed to construct the tile and then draw the tiling with a single command. These packages include many different methods for constructing tiles.

branching OUT

(a) Periodic tiling.　　　　　　　　　(b)　　　　　　　　(c) Aperiodic tilings.

Most of the tilings we consider in this text are *periodic* tilings. To explain what we mean by periodic, consider any tiling and imagine a transparent copy of the same tiling lined up exactly over the original tiling. The tiling is periodic if we can translate the copy—without any rotation—in at least two non-parallel directions so that it again lines up exactly over the original tiling. A tiling that is not periodic is called *aperiodic*. All of the regular and semiregular tilings are periodic as is the tiling in figure (a). The tilings shown in figures (b) and (c) are both aperiodic. In figure (b), the pattern cannot be translated back onto itself in any direction (although it can be rotated back onto itself), whereas in figure (c) the pattern can be translated horizontally, but not in a second direction.

It was at one time believed that if a set of one or more tiles could be used to construct an aperiodic tiling, then the same set of tiles could be used to construct a periodic tiling. However, in 1964, the logician Hao Wang discovered a set of over 20,000 tiles for which this assertion failed. Later, smaller sets of tiles with the same property were found, and finally in 1975 Roger Penrose, a mathematical physicist at Oxford, found a set containing only two tiles that could tile the plane nonperiodically but not periodically. The two tiles Penrose used, called kites and darts, are cut out from a special quadrilateral as shown in figure (d).

The angles of each of the tiles are labeled with an H for head or T for tail, and the rule for fitting the two Penrose tiles together is that only angles with the same letter may meet. If this rule were not in effect, then the tiles could be fitted together as they are in the quadrilateral and the plane could then be periodically tiled with this quadrilateral.

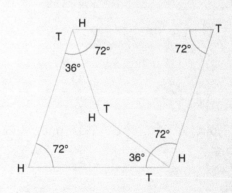

(d) Penrose kite and dart.

To avoid the necessity of imposing the head and tail rule for matching, the kites and darts can be modified so that the head and tail rule is forced by the shapes of the tiles, however, the Penrose tiling is usually presented in the kites and darts version with the head and tail rule in force. This Penrose tiling is shown in figure (e). In recognition of his wide-ranging service to science, Roger Penrose was knighted in 1994.

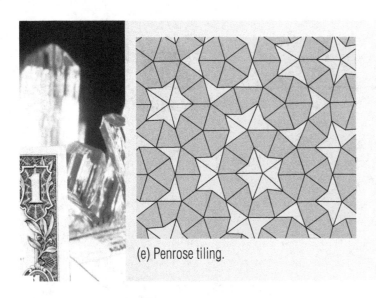

(e) Penrose tiling.

In a rather peculiar turn of events, Sir Roger Penrose noticed one day that the pattern on a roll of Kleenex quilted toilet paper strongly resembled one of his own nonperiodic patterns. In 1997, he and Pentaplex Ltd., which markets his patterns, filed suit against Kimberly-Clark Ltd. for copyright infringement. The suit seems to have been quietly settled.

We showed only two of the methods that can be used to make Escher-like tilings. We have only touched on the possibilities.

EXERCISES FOR SECTION 4.2

In Exercises 1 and 2, explain why the given tiling is not a regular tiling.

1.

2.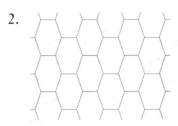

In Exercises 3–6, give at least one reason why the given tiling is not a semiregular tiling.

3.

4.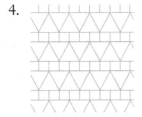

5.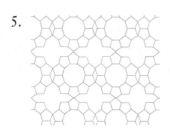

6.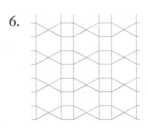

In Exercises 7–10, state the number of different vertex types in the given tiling and draw a sketch of each.

7.
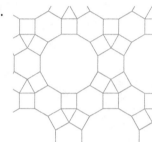

8.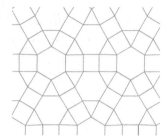
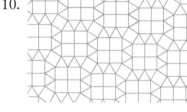

9.

10.

11. Find two different semiregular tilings that have two squares and three triangles about each vertex.

12. Find a semiregular tiling that does not include squares and that has four tiles about each vertex.

13. Find a semiregular tiling that does not include squares and that has five tiles about each vertex.

14. (a) Explain why it is not possible to construct a semiregular tiling where each vertex is of the type shown here.

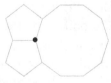

   (b) Find a semiregular tiling with two squares, an equilateral triangle, and a regular hexagon at each vertex. (The vertices have to be of a type different from the one shown in part (a).)

15. Explain why it is not possible to construct a semiregular tiling where each vertex is of the type shown here.

16. There is a vertex type that includes a regular 18-gon and two other regular polygons such that a semiregular tiling with the vertices of this type is not possible. Find the other two regular polygons in this vertex type.

17. There is a vertex type that includes a regular 24-gon and two other regular polygons such that a semiregular tiling with the vertices of this type is not possible. Find the other two regular polygons in this vertex type.

In Exercises 18–23, tile a sufficiently large region of the plane to illustrate how the given polygon can be used to tile the plane.

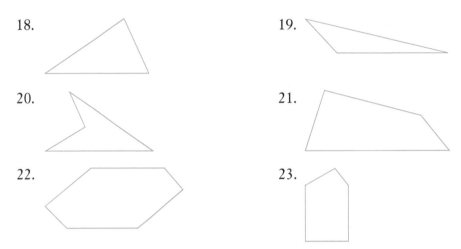

18.

19.

20.

21.

22.

23.

24. Suppose a rectangle has two sides modified as shown in the figure.

 (a) Modify the remaining two sides by the method illustrated in Figure 4.35.
 (b) Use the modified tile from part (a) to form a tiling.

25. Suppose a triangle has one half of one of its sides modified as shown in the figure.

 (a) Modify the other half of the side by the method illustrated in Figure 4.37, leaving the other sides unchanged.
 (b) Use the modified tile from part (a) to form a tiling.

26. Start with any rectangle and modify it by the method illustrated in Figure 4.35. Use this modified tile to form a tiling.

27. Start with any triangle and modify it by the method illustrated in Figure 4.37. Use this modified tile to form a tiling.

28. Using only regular polygon tiles, draw a tiling, which is not regular or semiregular, that has a pattern that can be extended to the entire plane.

In Exercises 29–32, tile a sufficiently large region of the plane to illustrate how the given polygon can be used to tile the plane. We have not looked at a systematic approach for doing this, so you will have to use trial and error.

29. **30.**

31. **32.**

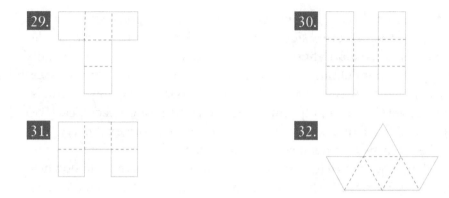

33. Consider the tiles shown in the figures. By flipping over the white tile, we can get the tile that is shaded blue. The definition of monohedral tiling allows for this kind of flipping of the tiles. Find at least two different tilings of the plane by this triangular shape, with some of the tiles face up and some flipped over. Tile a sufficiently large region to illustrate how the tiling can be extended to the entire plane.

34. Explain why there cannot be more than five regular polygons around a vertex in a semiregular tiling.

4.3 POLYHEDRA

Having considered the kinds of patterns that arise when tiling the plane with polygons, we will now look at the kinds of patterns that occur when polygons are used to form the surface of a solid figure. Such shapes and patterns can be seen in nature and in man-made objects; for instance, in the beautiful garnet crystals in Figure 4.39, in the Great Pyramid of ancient Egypt in Figure 4.40, and in the geodesic dome in Figure 4.41.

FIGURE 4.39 Garnet crystals.

FIGURE 4.40 The Great Pyramid.

FIGURE 4.41 Geodesic dome.

A solid figure, whose surface does not intersect itself, with polygonal sides, called **faces**, that are arranged so that every side of each face coincides entirely with a side of a bordering face is called a **polyhedron**. (The plural of polyhedron is polyhedra.) We further restrict our definition of polyhedron to prohibit cases of two faces sharing more than one side. A line forming the side of two faces of a polyhedron is called an **edge**. The **vertices** of a polyhedron are the vertices of the faces, each of which belongs to more than one face. A polyhedron is called **convex** if every line segment drawn between two points inside the polyhedron lies completely inside the polyhedron. A polyhedron that is not convex is called **concave**. The polyhedron in Figure 4.42(a) is convex, whereas the ones in Figures 4.42(b) and 4.42(c) are concave.

(a)   (b)   (c)

FIGURE 4.42 Polyhedra.

There is a relationship between the number of vertices, edges, and faces of a convex polyhedron. To discover this relationship ourselves, let's look at the few simple cases listed in Table 4.2. We use the letters V, E, and F to denote the number of vertices, edges, and faces, respectively. The first case is a cube, and it has $V = 8$, $E = 12$, and $F = 6$. You should verify that the number of vertices, edges, and faces for the other three convex polyhedra in Table 4.2 are correct.

TABLE 4.2 Vertices, Edges, and Faces of Some Polyhedra

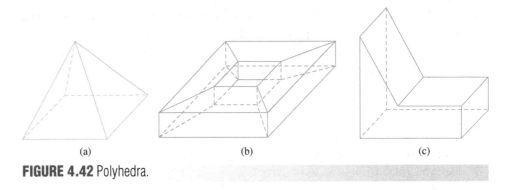

Polyhedron	Vertices: V	Edges: E	Faces: F
	8	12	6
	6	9	5
	10	15	7
	8	13	7

By looking at the numbers in Table 4.2, perhaps you already see a relationship between the number of vertices, edges, and faces. If not, try some more polyhedra on your own. After looking at many cases and searching for a relationship between V, E, and F, you will most likely happen upon the same formula discovered by the great mathematician Leonhard Euler:

Euler's Formula for Convex Polyhedra

If V is the number of vertices, E the number of edges, and F the number of faces of a convex polyhedron, then

$$V - E + F = 2$$

Euler's formula is valid for all polyhedra, convex or concave, as long as they do not have any "holes" going through them. Polyhedra without holes are called *simple polyhedra*. The polyhedra in Figures 4.42(a) and 4.42(c) are simple, whereas the polyhedron in Figure 4.42(b) has a hole through it and therefore is not simple.

EXAMPLE 1 *Verifying Euler's Formula*

Verify Euler's formula for the polyhedron in Figure 4.43.

Leonard Euler (1707–1783)

SOLUTION:

The polyhedron shown has 7 vertices, 11 edges, and 6 faces, so we have $V = 7$, $E = 11$, and $F = 6$. Therefore,

$$V - E + F = 7 - 11 + 6 = 2,$$

and Euler's formula is satisfied.

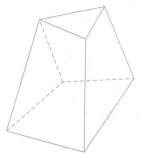

FIGURE 4.43

■

EXAMPLE 2 *Using Euler's Formula*

If a convex polyhedron has 14 vertices and 21 edges, how many faces does it have?

Substituting $V = 14$ and $E = 21$ into Euler's formula $V - E + F = 2$, we get

$$14 - 21 + F = 2$$
$$-7 + F = 2$$
$$F = 9,$$

so the polyhedron has 9 faces. ∎

EXAMPLE 3 *Finding an Unknown Polyhedron*

Suppose a convex polyhedron has 7 faces and 6 vertices. How many triangles, quadrilaterals, pentagons, and so on make up the faces of such a polyhedron? Sketch one polyhedron of this type.

Substituting $F = 7$ and $V = 6$ into Euler's formula, we get

$$6 - E + 7 = 2$$
$$13 - E = 2$$
$$E = 11,$$

FIGURE 4.44

so the polyhedron has 11 edges. Because each of the 11 edges belongs to exactly 2 faces, the polygonal faces have a total of $11 \cdot 2 = 22$ sides altogether. Because each of the 7 polygonal faces must have at least 3 sides, we see that the only way the total number of sides can add up to 22 is to have 6 triangular faces and 1 quadrilateral face. A sketch of one convex polyhedron with 6 triangular faces, 1 quadrilateral face, 11 edges, and 6 vertices is shown in Figure 4.44. ∎

EXAMPLE 4 *Finding an Unknown Polyhedron*

Suppose a convex polyhedron has 9 edges and 5 faces. Determine the number of sides each face must have and sketch one polyhedron of this type.

SOLUTION:

Letting $E = 9$ and $F = 5$ in Euler's formula, we see that

$$V - 9 + 5 = 2$$
$$V - 4 = 2$$
$$V = 6,$$

so the polyhedron has 6 vertices. Because the polyhedron has 9 edges, the polygonal faces have $9 \cdot 2 = 18$ sides altogether. Each of the 5 faces must have at least 3 sides, but this leaves three more sides. There are three possible combinations of 5 faces for which the total number of sides is 18:

1. 4 triangles and 1 hexagon,

2. 3 triangles, 1 quadrilateral, and 1 pentagon,

3. 2 triangles and 3 quadrilaterals.

Will 4 triangles and 1 hexagon work? Let's see. The hexagonal face would have 6 vertices and attaching one triangular face would have to contribute at least one more vertex, bringing the total number of vertices to be at least 7. But there are only supposed to be 6 vertices, so 4 triangles and 1 hexagon will not work. What about the second possibility of 3 triangles, 1 quadrilateral, and 1 pentagon? The pentagonal face would have 5 vertices and attaching the quadrilateral face would have to contribute at least two more vertices, bringing the total number of vertices to at least 7, so this will not work either. In the third case, with 2 triangle faces and 3 quadrilateral faces, we do not run into this kind of problem. One polyhedron of this type is shown in Figure 4.45.

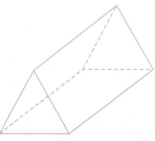

FIGURE 4.45

■

Regular Polyhedra

Which polyhedra are the most symmetric? A cube is very symmetric. As we see in Figure 4.46, all of the faces of a cube are regular polygons of the same size and shape and three edges meet at every vertex. The view from any face of a cube is the same as the view from any other face. In general, the most symmetric kind of polyhedra are the regular polyhedra, defined as follows:

FIGURE 4.46

A **regular polyhedron** is a convex polyhedron with all of its faces regular polygons of the same size and shape and with the same number of edges meeting at each vertex.

A cube is an example of a regular polyhedron, but there are others. The regular polyhedra were studied by the ancient Greeks, and the Greek mathematician Theaetetus (414–369 B.C.) proved that there are only five possible regular polyhedra. This proof is included as the final result in Euclid's *Elements* (circa 350 B.C.). A list of these five solids along with the shape of each face and the number of faces is given in Table 4.3, and the regular solids are shown in Figure 4.47.

TABLE 4.3 The Five Regular Polyhedra

Name	Shape of the Faces	Number of Faces
Tetrahedron	Equilateral triangles	4
Cube	Squares	6
Octahedron	Equilateral triangles	8
Dodecahedron	Regular pentagons	12
Icosahedron	Equilateral triangles	20

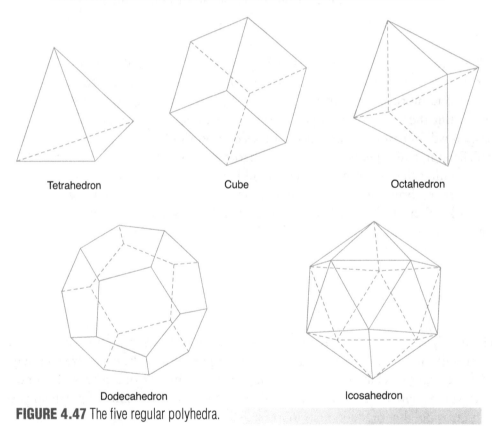

| Tetrahedron | Cube | Octahedron |

| Dodecahedron | Icosahedron |

FIGURE 4.47 The five regular polyhedra.

Notice that each of the regular polyhedra in Figure 4.47 is very symmetrical. One way to visualize this symmetry is to notice that if the polyhedron is rotated about its center so that one face is moved into the previous position of any other face, then the polyhedron would look exactly the same.

The five polyhedra in Figure 4.47 are the only regular polyhedra. The fact that no others exist can be proven by using Euler's formula. To see this, suppose we have a regular polyhedron whose faces are regular n-gons, and such that r edges meet at each vertex. Let V, E, and F be the number of vertices, edges, and faces of the polyhedron. Because each face has n edges and each edge belongs to two faces, we see that $nF = 2E$. Solving for F, we have

$$F = \frac{2E}{n}.$$

Similarly, because each vertex has r edges meeting it and each edge belongs to two vertices, we have $rV = 2E$, and therefore

$$V = \frac{2E}{r}.$$

Now, substituting these equations for F and V into Euler's formula $V - E + F = 2$, we get

$$\frac{2E}{r} - E + \frac{2E}{n} = 2.$$

Solving for E in this equation, we have

$$\left(\frac{2}{r} - 1 + \frac{2}{n}\right)E = 2,$$

$$E = \frac{2}{\left(\frac{2}{r} - 1 + \frac{2}{n}\right)}$$

We now have an equation for the total number of edges, E, in terms of the number of edges, r, meeting at each vertex and the number of edges on each face, n. Consider the faces around a particular vertex and imagine taking these faces and flattening them out, as illustrated for a vertex of a cube in Figure 4.48. In general, the sum of the angles of the r faces of the regular n-gons at a vertex must be less than 360°. Because r is at least 3, the angle of each n-gon face must be less than 360/3 = 120° the measure of the angle of a regular hexagon. Therefore, the n-gon faces have at most 5 sides, and we can conclude that $n = 3, 4$, or 5. Furthermore, because the angle of any regular n-gon is at least 60°, we know that there are fewer than 360°/60° = 6 n-gons meeting at each vertex, and therefore $r = 3, 4$, or 5. Trying the nine possible combinations of $n = 3, 4$, or 5 and $r = 3, 4$, or 5, the formula

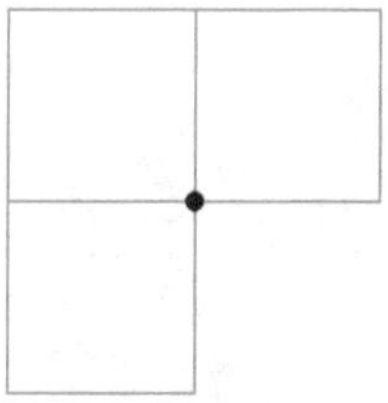

FIGURE 4.48

$$E = \frac{2}{\left(\frac{2}{r} - 1 + \frac{2}{n}\right)}$$

gives a positive integer value for E only in the five cases: $n = 3, r = 3$; $n = 3, r = 4$; $n = 3$, $r = 5$; $n = 4, r = 3$; and $n = 5, r = 3$. (We will leave the verification of this as an exercise.) But these are precisely the cases of the tetrahedron, octahedron, icosahedron, cube, and dodecahedron, respectively. Therefore, the five regular solids we have seen are the only ones possible.

The five regular polyhedra are sometimes called the **Platonic solids** because Plato (429–348 B.C.) attempted to relate the four basic elements of fire, earth, air, and water, plus the universe as a whole, to regular polyhedra. Plato was not the last to attempt to find deep significance in the fact that there are only five regular polyhedra. Johannes

Kepler (1571–1630), who first correctly described the elliptical orbits of the planets, originally proposed a rather peculiar theory of the solar system involving the regular solids. According to this theory, by circumscribing a dodecahedron about the sphere containing the orbit of Earth, and then circumscribing a sphere about the dodecahedron, the sphere containing the orbit of Mars can be found. Next, if a tetrahedron is circumscribed about the orbit of Mars, then the sphere surrounding this tetrahedron contains the orbit of Jupiter. Similarly, the sphere of the orbit of Venus lies inside the icosahedron inscribed in the sphere containing the orbit of Earth, and finally, the sphere orbit of Mercury lies inside the octahedron inscribed in the sphere containing the orbit of Venus. Kepler's own drawing illustrating this theory is shown in Figure 4.49.

© Bettmann/CORBIS

FIGURE 4.49 Kepler's solar system model.

These ideas of Plato and Kepler were a bit off the mark, but regular polyhedra do arise in nature. Some crystals grow in the shape of regular polyhedra because of the arrangement of atoms within them—for instance, the octahedral fluorite crystal in Figure 4.50. Many viruses have an icosahedral shape, for example the adenovirus in Figure 4.51 which is responsible many respiratory illnesses in humans.

Image © Fribus Ekaterina, 2012. Shutterstock, Inc.

FIGURE 4.50 Fluorite crystal.

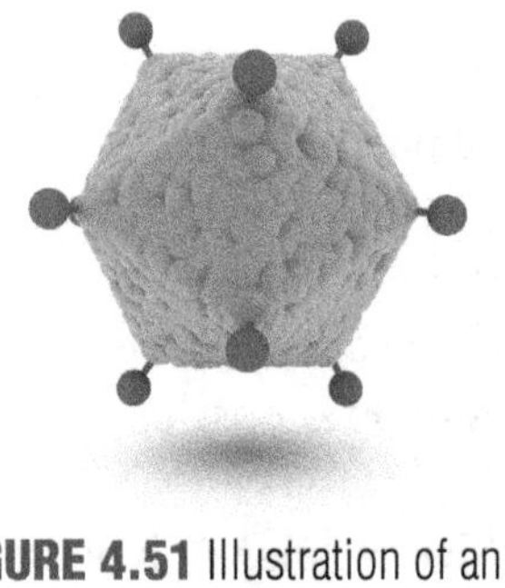

Image © Sebastian Tomus, 2012. Shutterstock, Inc.

FIGURE 4.51 Illustration of an adenovirus showing its icosahedron shape.

branching OUT

HOW MANY SIDES CAN A FAIR DIE HAVE?

Common dice are cubes. Because of its symmetry, a cube makes a fair die. When rolled, all six faces are equally likely to come out on the top. Similarly, fair dice can also be constructed in the shapes of the other four regular polyhedra. However, because a tetrahedron will land with a vertex up rather than a face up, we consider the face that is on the bottom to be the face that was rolled. These kinds of unusual dice can be found in some games. The game of Scattergories™ uses the icosahedral die shown in figure (a). The 20 faces are labeled with the letters of the alphabet (excluding Q, U, V, X, Y, and Z).

(a) Die from the Scattergories™ game.

(b) 10-sided fair die.

What about fair dice of other shapes? In figure (b) we see a fair die with 10 sides. This and the regular polyhedra are the most common dice in adventure games such as Dungeons and Dragons™ and Star Trek. This 10-sided die is symmetric in that all its faces are of the same size and shape and have identical surroundings. If the die were unlabeled, each of the faces would be indistinguishable from the others. For what number of faces is it possible to construct such a symmetrical die? It is easy to construct one when the number of faces desired is an even number that is at least six. To build such a die with $2n$ faces, where $n \geq 3$, simply begin with a regular n-gon and build a pyramid with identically shaped, triangular faces above and below it. The die in figure (b) is constructed slightly differently, but is in the same spirit. It turns out that constructing a symmetrical die with an odd number of faces is impossible.

Is it possible to construct fair dice that are not so symmetrical? With a die whose faces have distinct shapes, an analysis of the physics of the die and surface and of the manner in which it is rolled would allow computation of the probability of landing on each of the faces, but such an analysis is not practical to carry out. Of course, a very large number of rolls of any particular die would likely approximate the true probabilities quite well. However, it is very unlikely that by good fortune one could happen upon a nonsymmetrical die all of whose faces are equally likely to be rolled.

Semiregular Polyhedra

With tilings, we saw that there were only three regular tilings and eight semiregular tilings of the plane. Because there are only five regular polyhedra, it is natural to ask how many semiregular polyhedra are possible. A semiregular polyhedron is defined as follows:

A **semiregular polyhedron** is a polyhedron with all of its faces regular polygons, all of its vertices of the same type, with at least two different kinds of regular polygons as faces, and an additional symmetry condition that, up to mirror image, the polyhedron looks exactly the same from every vertex.

As in the case of tilings, we consider two vertices to be of the same **vertex type** if the same number and kind of polygons are arranged about them in the same order, except that one arrangement is possibly clockwise whereas the other is counterclockwise. Two infinite families of semiregular polyhedra—the prisms and antiprisms—are illustrated in Figure 4.52. **Prisms** have two regular n-gons for top and bottom, aligned with one directly above the other, and n square sides. **Antiprisms** again have two regular n-gons for top and bottom, with the top twisted by $180/n$ degrees, so that $2n$ equilateral triangles make up the sides.

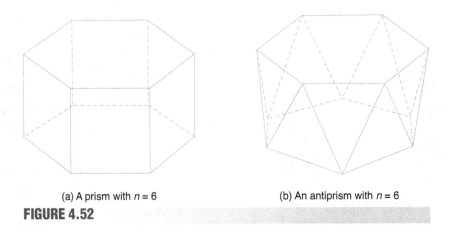

(a) A prism with $n = 6$ (b) An antiprism with $n = 6$

FIGURE 4.52

There are only 13 other possible convex semiregular polyhedra, which are sometimes called the **Archimedean solids** after the Greek scholar Archimedes (287–212 B.C.). (Two of the Archimedean solids have mirror images that are different from themselves. Each of these mirror image pairs has been counted as a single Archimedean solid.) Four of the Archimedean solids are illustrated in Figure 4.53.

As required, for each of the semiregular polyhedra shown in Figures 4.52 and 4.53, all vertices of the polyhedron are of the same type. For instance, the vertices of the rhombicuboctahedron in Figure 4.53(c) all have three squares and one equilateral triangle surrounding them. Notice that the angles of these four polygons add up to $90° + 90° + 90° + 60° = 330°$ and not $360°$, as would the angles at a vertex on a flat surface.

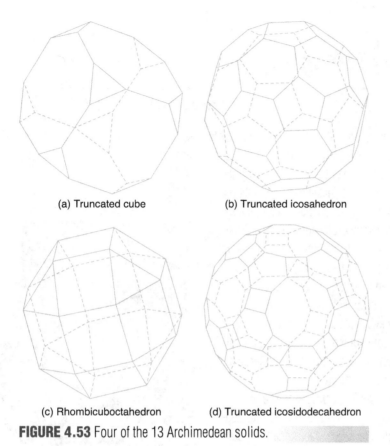

(a) Truncated cube

(b) Truncated icosahedron

(c) Rhombicuboctahedron

(d) Truncated icosidodecahedron

FIGURE 4.53 Four of the 13 Archimedean solids.

Seven of the 13 semiregular polyhedra can be obtained by starting with a regular polyhedron and successively slicing off (truncating) a vertex or edge. For instance, the truncated cube in Figure 4.53(a) is just a cube whose corners have been sliced off as shown in Figure 4.54 so that the resulting faces are regular octagons and equilateral triangles.

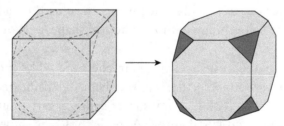

FIGURE 4.54 Constructing a truncated cube.

There is one more polyhedron whose faces are regular polygons and whose vertices are of the same type, but it does not satisfy the symmetry condition that would classify it as a semiregular polyhedron. This polyhedron is called the pseudo rhombicuboctahedron, and it may be formed by twisting the bottom third of the rhombicuboctahedron in Figure 4.53(c) by 45°.

branching OUT

TEXAS POLITICIANS THINK VERY SMALL

New York has its official state muffin, the apple muffin, Arizona has the bolo tie as its official state neckwear, and Ohio has honored tomato juice by naming it the official state beverage. But only Texas can lay claim to an official state molecule. After Rice University professors Richard Smalley and Robert Curl won the Nobel Prize in chemistry for their work on C_{60}, the buckyball, state Representative Scott Hochberg, a 1975 Rice alumnus, proposed legislation that would name the buckyball the state molecule of Texas. In spite of some competition from a molecule called Texaphyrin, which was engineered by chemists at the University of Texas, the buckyball proposal passed in 1997. At the same time, the Texas sweet onion became the state vegetable and picante sauce was named the state sauce. Texas has more state symbols than any other state. As explained by University of Texas Professor Janice May, "It's the second-biggest state, but first in braggarts." Nevertheless, Texas has yet to name an official state muffin.

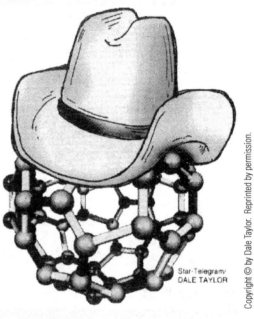

Star-Telegram/
DALE TAYLOR

FIGURE 4.55 A regulation soccer ball.

If the truncated icosahedron in Figure 4.53(b) looks familiar, that is probably because the pattern in this solid is the same pattern used on a regulation soccer ball, although the ball is inflated into a spherical shape, as shown in Figure 4.55.

The truncated icosahedron also appears in chemistry. The molecular structure of C_{60}, a form of carbon molecule, is in the shape of a truncated icosahedron. In these molecules, which look like tiny soccer balls, 60 carbon atoms lie at the 60 vertices of the truncated icosahedron. A model of C_{60} is shown in Figure 4.56, where the edges between the atoms represent the bonds between the atoms. The molecule C_{60} was discovered by chemists Richard Smalley and Robert Curl in 1985, and in 1996 they were awarded the Nobel Prize in chemistry for this work. The C_{60} molecule is called a *buckminsterfullerene* because it resembles the geodesic domes found in the designs of the architect R. Buckminster Fuller (1895–1983). These molecules are commonly called *buckyballs*. The buckyball is one of a family of carbon molecules called *fullerenes*. In a fullerene, the carbon atoms and their bonds form a polyhedron with hexagonal and pentagonal faces. Each carbon atom is joined to three others. Using these facts, combined with Euler's formula, it can be shown that any fullerene must have exactly 12 pentagonal faces. We leave the verification of this as an exercise. A buckyball has 12 pentagonal faces and 20 hexagonal faces. Since the discovery of the buckyball, other fullerenes have been found. One of them, C_{70}, has an oblong shape, and it has 12 pentagonal and 25 hexagonal faces. Another, C_{44}, is a smaller fullerene made up of 12 pentagons and 12 hexagons. There are also very large fullerenes such as C_{540}. This certainly invites the question of what fullerenes are geometrically possible given the limitation of having only pentagonal and hexagonal faces and three edges at each vertex. This question was answered by mathematicians Branko Grünbaum and Theodore S. Motzkin in 1963, more than 30 years before the discovery of fullerenes. Their results showed that there cannot be a fullerene with exactly one hexagonal face, but any other number of hexagonal faces, including possibly none, is geometrically possible.

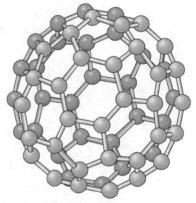

FIGURE 4.56 The buckyball molecule, C_{60}.

EXERCISES   FOR SECTION 4.3

In Exercises 1–4, determine whether the polyhedron is convex or concave.

1.
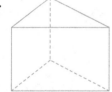

2.
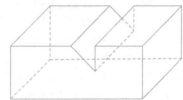

3.
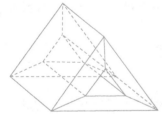

4.
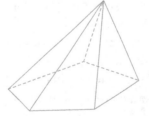

In Exercises 5–8, verify Euler's formula for the polyhedron.

5.
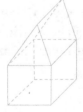

6.

7.

8.

In Exercises 9–13, find the number of vertices, edges, and faces and verify Euler's formula for the given regular polyhedron.

9. tetrahedron   10. cube
11. octahedron   12. dodecahedron
13. icosahedron

14. If a convex polyhedron has 25 edges and 12 faces, how many vertices does it have?
15. If a convex polyhedron has 20 edges and 10 vertices, how many faces does it have?
16. If a convex polyhedron has 33 faces and 33 vertices, how many edges does it have?
17. Suppose a convex polyhedron has six faces and five vertices. How many triangles, quadrilaterals, pentagons, and so on make up the faces of such a polyhedron? Sketch one polyhedron of this type.
18. Suppose a convex polyhedron has 13 edges and 8 vertices and all of its faces are triangles and quadrilaterals. Find the number of triangles and quadrilaterals making up the faces of such a polyhedron and sketch one polyhedron of this type.
19. Suppose a convex polyhedron has 10 edges and 6 vertices. How many triangles, quadrilaterals, pentagons, and so on might make up the faces of such a polyhedron? (There are two possible answers.) Sketch an example of a polyhedron of each of these two types.
20. Suppose a convex polyhedron has 7 faces and 10 vertices and none of its faces are triangles. How many quadrilaterals, pentagons, hexagons, and so on make up the faces of such a polyhedron? Sketch one polyhedron of this type.
21. Suppose a convex polyhedron has eight faces, of which four are triangles and four are quadrilaterals.
    (a) How many edges does the polyhedron have?
    (b) How many vertices does the polyhedron have?
    (c) Sketch one polyhedron of this type.
22. Suppose a convex polyhedron has 12 faces, of which 4 are triangles, 7 are quadrilaterals, and 1 is a hexagon.
    (a) How many edges does the polyhedron have?
    (b) How many vertices does the polyhedron have?

Euler's formula holds for all polyhedra, convex or concave, as long as they do not have any "holes" through them as in the polyhedron in Figure 4.42(b). Polyhedra without holes are called simple polyhedra. In Exercises 23 and 24, verify that Euler's formula holds for the simple concave polyhedron shown.

23.
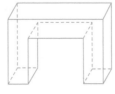

24.
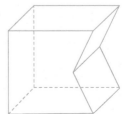

In Exercises 25–28, find the number of vertices, edges, and faces of the polyhedron.

25.

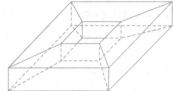

26.

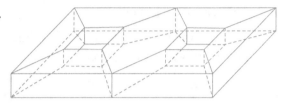

27.

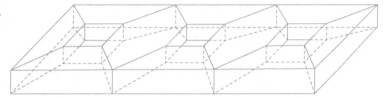

28.

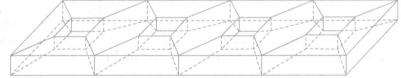

29. There is a formula similar to Euler's formula that relates the number of vertices, edges, and faces of a polyhedron with *n* holes. Based on the answers to Exercises 25–28, try to find this formula.

30. A cuboctahedron is a semiregular solid that can be constructed by slicing off the corners of a cube through the midpoints of the edges as shown here.

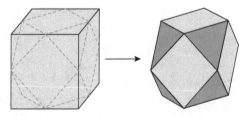

(a) How many triangular faces does a cuboctahedron have?
(b) How many square faces does a cuboctahedron have?

31. A truncated octahedron is a semiregular solid that can be constructed by slicing off the corners of an octahedron so that the resulting faces are regular hexagons and squares as shown in the figure.

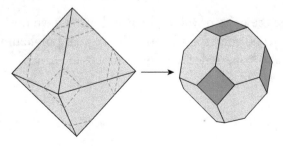

    (a)   How many hexagonal faces does a truncated octahedron have?

    (b)   How many square faces does a truncated octahedron have?

32. The cuboctahedron, constructed in Exercise 30 by slicing off the corners of a cube, can also be constructed by slicing off the corners of another regular solid. Which regular solid is this?

33. A truncated tetrahedron is a semiregular solid formed by slicing off the corners of a tetrahedron. What types of regular polygons appear as faces of a truncated tetrahedron, and how many faces are there of each type?

If a cube is drawn and the centers of each face of the cube are connected by lines, an octahedron lying within the cube is formed as shown in the figure.

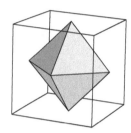

We say that the octahedron is the dual of the cube. In general, the dual of any regular polyhedron is another regular polyhedron formed by connecting the centers of the faces of the original polyhedron. In Exercises 34–37, find the dual of the given regular polyhedron.

34. tetrahedron   35. octahedron

36. dodecahedron   37. icosahedron

38. In the text we saw that if a regular polyhedron has E edges, faces that are regular n-gons, and r edges meeting at each vertex, then the following equation is satisfied:

$$E = \frac{2}{\left(\frac{2}{r} - 1 + \frac{2}{n}\right)}$$

We also saw that $n = 3$, 4, or 5 and $r = 3$, 4, or 5.

    (a)   In the following table, use this formula for E to fill in the value of E for each of the nine possible pairs of values for n and r.

n	3	3	3	4	4	4	5	5	5
r	3	4	5	3	4	5	3	4	5
E									

    (b)   Show how the pairs of values for n and r that gave positive integer values for E in part (a) correspond to the five regular polyhedra: the tetrahedron, cube, octahedron, dodecahedron, and icosahedron.

39. Suppose a fullerene has p pentagonal faces, h hexagonal faces, and a carbon atom at each vertex. Each carbon atom is joined by three bonds to three other carbon atoms, and these bonds form the edges of a polyhedron.
    (a) Explain why the fullerene has $\frac{1}{3}(5p + 6h)$ vertices.
    (b) Explain why the fullerene has $\frac{1}{2}(5p + 6h)$ edges.
    (c) Use the results of parts (a) and (b) and Euler's formula to show that any fullerene must have exactly 12 pentagonal faces.
40. Use the results of Exercise 39 to find the number of hexagonal faces in the fullerene C_{540}.
41. Show that for any polyhedron with V vertices, E edges, and F faces,
    (a) $E \geq \frac{3}{2}F$,
    (b) $E \geq \frac{3}{2}V$.
42. Using Euler's formula and the results of Exercise 41, find all of the pairs of values of E and V for which there may be a convex polyhedron with seven faces.
43. Using Euler's formula and the results of Exercise 41, find all of the pairs of values of E and V for which there may be a convex polyhedron with eight faces.
44. Show that for any polyhedron, the number of faces that have an odd number of sides is even.

WRITING EXERCISES

1. Explain some of the practical advantages of using an edge-to-edge tiling when tiling a floor and the possible disadvantages of laying a brick wall in an edge-to-edge pattern.
2. Discuss some of the practical advantages of using square tiles instead of tiles of some other shape.
3. Explain why you think cubical dice seem to be the most common randomizing device for board games, rather than a die in the shape of some other regular polyhedron or a different type of randomizing device such as a spinner or drawing chips with numbers.
4. A die in the shape of a pyramid with a square base and four triangular faces with a common vertex above the center of the base may be tall and thin or short and wide. Explain why adjusting these relative proportions should allow one to construct a fair die.

PROJECTS

1. We saw that a quadrilateral of any shape can be used in a monohedral tiling of the plane. In our construction, every vertex contains all four angles of the quadrilateral. For certain special quadrilaterals, another way of arranging the tiles yields a different monohedral tiling. Find some examples using rhombuses

with particular angles. (A *rhombus* is a quadrilateral with all four sides of equal length. It is necessarily a parallelogram, with opposite sides parallel.)

2. A pentomino is a polygon built up from five unit squares. They must be connected through shared edges (not partial edges or just corners). Two such pentominoes are illustrated here.

   (a) Show that either of these pentominoes can be used in a monohedral tiling of the plane.
   (b) There are twelve different pentominoes (if you are allowed to flip them over), any of which tile the plane in a monohedral tiling. Find as many as you can and show that they tile the plane.

3. In addition to the translational symmetry exhibited by periodic tilings as described in Branching Out 4.3, many tilings also exhibit rotational symmetry. A tiling has rotational symmetry if it is possible to rotate the entire tiling about some point in the tiling so that the rotated tiling lines up exactly with the original tiling.

   (a) Which regular and semiregular tilings exhibit rotational symmetry? What are the points about which you can rotate and the angles through which you can rotate?
   (b) Can you find other tilings with rotational symmetry? Try to produce some different sets of angles through which you can rotate.

4. Find examples of tessellations and polyhedra that occur in nature. Comment on any structural or evolutionary advantages conferred by such patterns.

5. We can alter a hexagon by adding or subtracting identical small, equilateral triangles from certain edges, as in the three examples shown here.

   (a) Which of these shapes can be used in a monohedral tiling of the plane? Consider many different ways tiles could fit together. Use deductive reasoning to keep the number of cases manageable. Explain how you can eliminate certain cases.
   (b) Design some tiles built up in a similar way from hexagons, and determine whether they can be used to form a monohedral tiling of the plane.

6. The *angle defect* of a vertex of a polyhedron is defined to be 360° minus the sum of the face angles at the vertex. (The angle defect may be negative.) The *total angle defect* of a polyhedron is the sum of the angle defects over all vertices of a polyhedron.
   (a) Find the total angle defect for each of the regular polyhedra. What do you get?
   (b) Find the total angle defect for some other convex polyhedra. Make a conjecture about what can be said about the total angle defect for convex polyhedra.
   (c) Suppose a convex polyhedron has n_3 triangular faces, n_4 quadrilateral faces, n_5 pentagonal faces, and so forth. Express V, E, F, and the total angle defect in terms of $n_3, n_4, n_5, \ldots$. Try to use Euler's formula to prove your conjecture from part (b).
   (d) What happens in concave polyhedra or in polyhedra with holes such as those in Exercises 25–28 of Section 4.3?

KEY TERMS

angle of a polygon, 248	face of a polyhedron, 270	semiregular polyhedron, 278
antiprism, 278	monohedral tiling, 253	semiregular tiling, 257
Archimedean solid, 278	*n*-gon, 247	side of a polygon, 247
concave polygon, 247	Platonic solids, 275	tessellation, 253
concave polyhedron, 270	polygon, 246	tiling, 253
convex polygon, 247	polyhedron, 270	vertex of a polygon, 247
convex polyhedron, 270	prism, 278	vertex of a polyhedron, 270
edge of a polyhedron, 270	regular polygon, 248	vertex of a tiling, 254
edge-to-edge tiling, 253	regular polyhedron, 273	vertex type, 255, 278
Euler's formula, 271	regular tiling, 254	

CHAPTER 4 / REVIEW TEST

1. Sketch a concave hexagon.

2. Sketch a convex 11-gon.

3. Find the measure of an angle of a regular 15-gon.

4. Five of the six angles of the following polygon are given. Find the remaining angle.

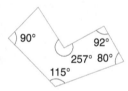

5. Tile a sufficiently large region of the plane to illustrate how the triangle shown here can be used to tile the plane.

6. Tile a sufficiently large region of the plane to illustrate how the quadrilateral shown here can be used to tile the plane.

7. State the number of different vertex types in the given tiling and draw a sketch of each one.

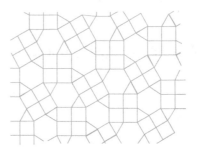

8. Find a semiregular tiling that does not include octagons and that has three tiles about each vertex. (There are two possible answers.)

9. There is a vertex type that includes a regular 42-gon and two other regular polygons such that a semiregular tiling with vertices of this type is not possible. Find the other two regular polygons in this vertex type.

10. Explain why it is not possible to construct a semiregular tiling where each vertex is of the type shown in the figure.

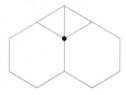

11. The tiling shown here from the coloring book *Altair Design, Book 1*, appears, at first glance, to be a tiling with regular polygon tiles. Explain why it is impossible for all of these polygons to be regular.

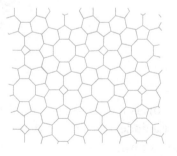

12. Verify Euler's formula for the following polyhedron.

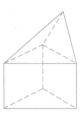

13. If a convex polyhedron has 14 faces and 13 vertices, how many edges does it have?

14. If a convex polyhedron has 44 edges and 22 faces, how many vertices does it have?

15. Suppose a convex polyhedron has eight edges and five faces. How many triangles, quadrilaterals, pentagons, and so on make up the faces of such a polyhedron? Sketch a picture of one polyhedron of this type.

16. A truncated dodecahedron is a semiregular solid that can be constructed by slicing off the corners of a dodecahedron so that the resulting faces are regular triangles and decagons as shown in the figure.

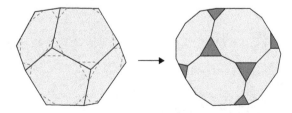

(a)   How many triangular faces does a truncated dodecahedron have?

(b)   How many decagonal faces does a truncated dodecahedron have?

(c)   Find the number of vertices, faces, and edges and verify Euler's formula for the truncated dodecahedron.

SUGGESTED READINGS

Danzer, Ludwig, Branko Grünbaum, and G. C. Shephard. "Equitransitive Tilings, or How to Discover New Mathematics." *Mathematics Magazine* 60:2 (April 1987), 67–88. An investigation of a class of tilings more general than semiregular tilings, written in a way that illustrates the process of mathematical research.

Gardner, Martin. "Extraordinary Nonperiodic Tiling That Enriches the Theory of Tiles." *Scientific American* 236:1 (January 1977), 110–121. Reprinted with additional updates in Martin Gardner, *Penrose Tiles to Trapdoor Ciphers*. New York: Freeman, 1989, 1–29. An introduction to Penrose tilings.

Gardner, Martin. "On Tessellating the Plane with Convex Polygon Tiles." *Scientific American* 233:1 (July 1975), 112–117, 132. Reprinted with an addendum in Martin Gardner, *Time Travel and Other Mathematical Bewilderments*. New York: Freeman, 1988, 163–176. The original article listed the known results about monohedral tilings in the plane with convex polygons, including R. B. Kershner's incorrect claim to have found all convex pentagon tilings. The addendum to the article presented new pentagon tilings found by Richard E. James III and Marjorie Rice.

Grünbaum, Branko, and G. C. Shephard. "Some Problems on Plane Tilings." In *The Mathematical Gardner*, edited by David A. Klarner. Boston: Prindle, Weber & Schmidt, 1981, 167–196. On testing a tile to see whether it can be used in a monohedral tiling. Heesch's problem.

Grünbaum, Branko, and G. C. Shephard. *Tilings and Patterns*. Mineola, New York: Dover, 2012. A comprehensive treatment of the subject. The standard text in the field.

Kepes, Gyorgy. *Module, Proportion, Symmetry, Rhythm*. New York: George Braziller, 1966. Patterns in nature, art, and architecture.

O'Daffer, Phares G., and Stanley R. Clemens. *Geometry: An Investigative Approach*. Reading, Massachusetts: Addison-Wesley, 1992. An introductory geometry textbook emphasizing exploration.

Rigby, J. F. "Napoleon, Escher, and Tessellations." *Mathematics Magazine* 64:4 (October 1991), 242–246. Proofs of Napoleon's Theorem and Escher's Theorem based on tessellations.

Schattschneider, Doris. "In Praise of Amateurs." In *The Mathematical Gardner*, edited by David A. Klarner. Boston: Prindle, Weber & Schmidt, 1981, 140–166. Monohedral tilings with pentagons, emphasizing the discoveries of amateur mathematician Marjorie Rice.

Schattschneider, Doris. "The Plane Symmetry Groups: Their Recognition and Notation." *The American Mathematical Monthly* 85:6 (June–July 1978), 439–450. Possible symmetries of tilings.

Schattschneider, Doris. "Tiling the Plane with Congruent Pentagons." *Mathematics Magazine* 51:1 (January 1978), 29–44. Monohedral tilings with pentagons.

Schattschneider, Doris. *Visions of Symmetry*. New York: Harold N. Abrams, 2004. An extensive treatment of the work of M. C. Escher, including mathematical explanations, pages from his notebooks, and over 350 illustrations.

Seymour, Dale, and Jill Britton. *Introduction to Tessellations*. Palo Alto, Calif.: Dale Seymour, 1989. A nontechnical introduction to tiling. The chapter on creating Escher-like tilings shows some other techniques that are not included in this text.

Stevens, Peter S. *Patterns in Nature*. Boston: Little, Brown, 1974. Includes tiling patterns and polyhedra that occur in nature.

Thompson, D'Arcy. *On Growth and Form: The Complete Revised Edition*. Mineola, New York: Dover, 1992. Covers honeycombs of bees and the skeletons of radiolaria.

Washburn, Dorothy K., and Donald W. Crowe, *Symmetries of Culture*. Seattle: University of Washington Press, 1988. Includes a chapter on tiling patterns found in various cultures.

Image © Nomad_Soul, 2012. Shutterstock, Inc.

Number theory is the mathematics of numbers, especially the integers, . . . , –3, –2, –1, 0, 1, 2, 3, ... In number theory, we look at divisibility of numbers, patterns in the integers, and solutions to equations. Until recently, number theory was almost exclusively considered to be "pure" mathematics. By this we mean mathematics that is studied for its own intrinsic beauty rather than for its applications. However, today, number theory has found its way into many practical applications. For instance, with modern electronic communications, encoding messages has become an area of vital importance in which number theory plays a central role. In this chapter, we begin by looking at some of the basic ideas of number theory and some fascinating unsolved problems. We then go on to study a special kind of arithmetic used in number theory, called modular arithmetic, and some of its applications.

NUMBER THEORY

CHAPTER 5

5.1 DIVISIBILITY AND PRIMES

The first numbers we encounter are those used in counting: 1, 2, 3, ... However, when we subtract these numbers from one another, we encounter negative numbers and zero. For this reason, we will consider the entire set of **integers**, . . . , –3, –2, –1, –0, 1, 2, 3, ... When given a particular integer such as the number 4, the first thing we might try to do with it is break it down into other positive integers. This can be done through addition—for instance, with 4 as

$$4 = 3 + 1 = 2 + 2 = 2 + 1 + 1 = 1 + 1 + 1 + 1.$$

This leads us to the major area of number theory called *additive number theory*. However, we will focus on writing integers as products of other positive or negative integers, or *multiplicative number theory*. For example, the number 4 can be written as

$$4 = 4 \cdot 1 = 2 \cdot 2 = (-4)(-1) = (-2)(-2).$$

In this case, we say that 1, –1, 2, –2, 4, and –4 are all factors of 4, and that 4 is divisible by each of these numbers. More generally, we have the following definitions:

> If a and b are integers and there is an integer c such that $a = b \cdot c$, then we say that b **divides** a or a is **divisible** by b, and write $b \,|\, a$. In this case, we say that b is a **factor** or **divisor** of a.

For example, because $18 = 6 \cdot 3$, we see that $6\,|\,18$ and $3\,|\,18$. Similarly, $18 = (-9)(-2)$, so $(-9)\,|\,18$ and $(-2)\,|\,18$. Because $18/5 = 3.6$ is not an integer, we see that 5 does not divide 18. Starting from 1 and –1, we can list all of the divisors of 18: 1, –1, 2, –2, 3, –3, 6, –6, 9, –9, 18, and –18. We note here that the number 0 is divisible by any integer b because $0 = b \cdot 0$.

We just saw that the number 18 has six positive divisors. There are some positive integers—for example, the number 7—with only two positive divisors. We have a special name for such integers.

> A **prime number** is an integer greater than 1 whose only positive divisors are 1 and itself. An integer greater than 1 that is not prime can be written as a product of integers greater than 1 and is called a **composite number**.

The first six prime numbers are 2, 3, 5, 7, 11, and 13. Prime numbers are the basic multiplicative building blocks of all integers. This fact, called the **Fundamental**

Theorem of Arithmetic, was known to the ancient Greeks. It can be stated formally as follows:

> ### The Fundamental Theorem of Arithmetic
> Every integer greater than 1 can be expressed as a product of primes, and this representation is unique up to the order in which the factors occur.

To find the prime factorization of a given composite integer, first factor the integer into any two positive factors. If the factors are primes, then we are done. If any of the factors are not prime, continue factoring any composite factors into two factors until all the numbers in the factorization are prime. This technique is illustrated in the following example:

EXAMPLE 1 *Prime Factorization*

Find the prime factorization of the number 3150.

SOLUTION:

Integers ending in a 0 are divisible by 10, so first we can factor 3150 as

$$3150 = 10 \cdot 315.$$

Now factoring the 10 into $2 \cdot 5$ and the 315 into $5 \cdot 63$, we have

$$3150 = 2 \cdot 5 \cdot 5 \cdot 63.$$

The factor 63 is the product of 7 and 9, and therefore is not prime. We can factor it to get

$$3150 = 2 \cdot 5 \cdot 5 \cdot 7 \cdot 9.$$

Finally, factoring the only remaining composite factor, 9, into $3 \cdot 3$ gives the prime factorization

$$3150 = 2 \cdot 5 \cdot 5 \cdot 7 \cdot 3 \cdot 3.$$

It is standard to write the prime factors from the smallest to the largest and to combine any powers of the same prime factor. Doing this for our factorization of 3150, we get

$$3150 = 2 \cdot 3^2 \cdot 5^2 \cdot 7.$$

Prime Numbers

Prime numbers have always fascinated professional and amateur mathematicians. There are many questions we might ask about prime numbers—some of these have been answered, whereas others remain unresolved. A reasonable question to begin with is "How many prime numbers are there?" This question was answered by the Greek mathematician Euclid (circa 350 B.C.) in his famous work, the *Elements*, in which he gives an elegant proof that there are an infinite number of primes.

Another question one might ask is how to go about finding the prime numbers. The Greek scholar Eratosthenes of Cyrene (276–194 B.C.) developed a technique known as the **Sieve of Eratosthenes** for finding all of the prime numbers less than or equal to some given number. To see how the sieve works, suppose we want to find all prime numbers less than or equal to 50. We first list all integers between 2 and 50. The first number, 2, is prime, so we draw a box around it to indicate this. Next, we cross out any of the remaining numbers that are divisible by 2, in other words all of the remaining even numbers 4, 6, 8, ... The next number after 2 that has not been crossed out is 3, which must be prime because it is not divisible by any smaller

A CLOSER *look*

HOW DO WE KNOW THERE ARE AN INFINITE NUMBER OF PRIMES?

The proof that there are an infinite number of primes dates back at least to the time of Euclid, and most likely was known even earlier. The very simple and convincing proof found in the *Elements* is based on the following idea: Suppose we take the first four prime numbers 2, 3, 5, and 7. Consider the number

$$2 \cdot 3 \cdot 5 \cdot 7 + 1 = 211.$$

We know that 211 is not divisible by 2, 3, 5, or 7, because when dividing by any of these numbers there would be a remainder of 1. Therefore, 211 must be either prime or divisible by some prime other than 2, 3, 5, or 7. In either case, we know of the existence of some prime other than the four we started with. We can use this idea to show that there are an infinite number of primes. Suppose we multiply the first k prime numbers, 2, 3, 5, . . ., p_k, together and add 1 to get

$$N = 2 \cdot 3 \cdot 5 \cdots p_k + 1.$$

We know that N is not divisible by any of the first k prime numbers, 2, 3, 5, . . ., p_k. Therefore, N is either prime itself or it is the product of some other primes that are not in our current list. In either case, we have found a new prime. Therefore, we see that if we list the first k primes, we can always find another bigger prime. From this we can conclude that there must be an infinite number of primes.

prime number. We now draw a box around 3 to indicate that it is prime, and then we proceed to cross out any of the other numbers that are divisible by 3—in other words, every third number 6, 9, 12, ... Some of these numbers already have been crossed out at this point. Continuing in this way, we see that after crossing out the multiples of 7 nothing more is crossed out. We arrive at the list in Figure 5.1 with all of the primes boxed and all of the composite numbers crossed out.

FIGURE 5.1 The Sieve of Eratosthenes.

The Sieve of Eratosthenes provides a way of finding all primes less than a certain number. However, it is fairly time consuming. What if we just want to know whether or not one particular positive integer n is prime? One way to tell would be to divide n by every smaller positive integer. However, this is more than we have to do, for if n is composite, then it can be written as $n = a \cdot b$ with a and b both integers greater than 1. It cannot be true that both of these numbers are larger than $\sqrt{n}$ for if a and b were both greater than $\sqrt{n}$ we would have

$$n = a \cdot b > \sqrt{n} \cdot \sqrt{n} = n.$$

Because this is not possible, we know that either a or b is no larger than $\sqrt{n}$ and this factor must have a prime factor no larger than $\sqrt{n}$. We have deduced the following test for primality:

> **Test for Primality**
>
> An integer n greater than 1 is prime if it is not divisible by any prime number in the range from 2 to $\sqrt{n}$, inclusive.

Of course, if you do not have a list of all prime numbers up to $\sqrt{n}$, then you can just check to see if n is divisible by any integer in the range from 2 and $\sqrt{n}$, inclusive. Note that from this test for primality it follows that, when finding all primes less than or equal to n with the Sieve of Eratosthenes, we need only cross out multiples of all primes up to $\sqrt{n}$.

EXAMPLE 2 *Primality Testing*

Determine whether or not the number 617 is prime.

Because $\sqrt{617} \approx 24.84$, we only have to see if 617 is divisible by primes in the range from 2 to 24.84. Looking at the primes from our Sieve of Eratosthenes in Figure 5.1, we see that we only have to check for divisibility by the primes 2, 3, 5, 7, 11, 13, 17, 19, and 23. Because 617 is odd, 2 does not divide 617. Because

$$617/3 \approx 205.67,$$

3 does not divide 617. Similarly,

$$617/5 = 123.4,$$

so 617 is not divisible by 5. (Or we could have seen that 617 is not divisible by 5 because 617 does not have last digit 5 or 0.) Similarly, we can show that 617 is not divisible by 7, 11, 13, 17, 19, or 23. Therefore, we conclude that 617 is prime. ∎

EXAMPLE 3 *Primality Testing*

Determine whether or not the number 259 is prime.

Because $\sqrt{259} \approx 16.09$, we only have to see if 259 is divisible by any of the primes 2, 3, 5, 7, 11, or 13. Because 259 is odd, it is not divisible by 2. Computing $259/3 \approx 86.33$ and $259/5 = 51.8$, we see that 259 is not divisible by 3 or 5. Next we compute

$$259/7 = 37,$$

and we see that $259 = 7 \cdot 37$, so 259 is not prime. ∎

Much more sophisticated tests for primality are available today. These tests can determine whether an integer with 200 digits is prime in a few seconds of computer time. These tests are not able to factor a number, but merely determine whether or not it is prime. You might ask why we should want to find large primes. As we will see in Sections 5.6 and 5.7 of this chapter, large primes play a key role in certain secret code systems, so there is now a great practical interest in finding primes that are several hundred digits long. However, the problem of finding the *largest* known prime, which as of early 2012 was the number $2^{43,112,609} - 1$, a number with 12,978,189 digits (about 4000 pages worth), is a challenge that will continue to be of interest even though there are no immediate applications for primes this large.

Throughout mathematical history, many have attempted to find quick and easy formulas that will generate prime numbers as far out as one would care to go. All of these attempts failed. One formula that gives primes for many values of n is

$$n^2 - n + 41.$$

For instance,

$$\text{When } n = 0, \quad 0^2 - 0 + 41 = 41 \text{ is prime,}$$
$$\text{When } n = 1, \quad 1^2 - 1 + 41 = 41 \text{ is prime,}$$
$$\text{When } n = 2, \quad 2^2 - 2 + 41 = 43 \text{ is prime, and}$$
$$\text{When } n = 3, \quad 3^2 - 3 + 41 = 47 \text{ is prime,}$$

The formula $n^2 - n + 41$ continues to generate primes until we get to $n = 41$, in which case we get

$$41^2 - 41 + 41 = 1681 = 41 \cdot 41.$$

Another formula that showed some promise of giving only primes was proposed by the famous French mathematician Pierre de Fermat (1601–1665). Fermat noted that for $n = 0, 1, 2, 3,$ and 4, the formula

$$F_n = 2^{2^n} + 1$$

gives only primes. For instance, when $n = 0$,

$$F_0 = 2^{2^0} + 1 = 2^1 + 1 = 3$$

is prime, and when $n = 1$,

$$F_1 = 2^{2^1} + 1 = 2^2 + 1 = 5$$

is prime. Fermat believed that the numbers F_n, which are now called **Fermat numbers**, were prime for all integers $n \geq 0$. Unfortunately, in 1732 he was proved wrong by another great mathematician, Leonhard Euler (1707–1783), who showed that

$$F_5 = 2^{2^5} + 1 = 4{,}294{,}967{,}297 = 641 \cdot 6{,}700{,}417.$$

Many other Fermat numbers have now been shown to be composite. So far, only the Fermat numbers F_n with $n = 0, 1, 2, 3,$ and 4 are known to be prime, and no one knows if there are any more prime Fermat numbers. It is known that F_n is composite for $n = 5$ through 32. Although no prime factors for F_{20} and F_{22} are known, primality tests have shown them to be composite. As of early 2012, the largest Fermat number known to be composite was $F_{2{,}543{,}528}$, and the prime factor $9 \cdot 2^{2{,}543{,}551} + 1$ was discovered in 2011. It is difficult to test Fermat numbers for primality because they get large so quickly. For instance, even F_{24} has over five million digits!

Another question concerning primes is whether or not there are infinitely many pairs of consecutive odd numbers that are primes. These pairs of primes are called **twin primes**. For example, 5 and 7 are twin primes, as are the pair 11 and 13. Although the twin prime pairs become more scarce as we look at larger and larger numbers, no one knows if they continue forever. As of early 2012, the largest known twin prime pair was the one found on December 25, 2011:

$$3{,}756{,}801{,}695{,}685 \cdot 2^{666{,}669} - 1 \quad \text{and} \quad 3{,}756{,}801{,}695{,}685 \cdot 2^{666{,}669} + 1.$$

These numbers have 200,700 digits. It is interesting to note that both the largest prime and the largest twin prime pair were discovered by members of large, world-wide groups searching for primes on individual personal computers.

Christian Goldbach (1690–1764) raised another question about primes in a 1742 letter to Euler in which he made a conjecture that can be generalized slightly to the statement, "Every even integer greater than 4 can be written as a sum of two odd prime numbers." This conjecture came to be known as **Goldbach's conjecture**. It is easy to show that Goldbach's conjecture is true for the first several even integers greater than 4:

$$6 = 3 + 3,$$
$$8 = 3 + 5,$$
$$10 = 3 + 7 = 5 + 5,$$
$$12 = 5 + 7,$$
$$14 = 3 + 11 = 7 + 7.$$

Notice that in some cases there is more than one way to write the integer as the sum of two odd prime numbers. Goldbach's conjecture is easy to verify on a computer for a particular even integer greater than 4, and it has been shown to hold for all even integers less than 1,609,000,000,000,000,000. However, it is not known if it holds for all even integers greater than 4, so it remains one of the most famous unsolved problems in number theory.

There are numerous other unanswered questions about primes. It is these mysteries that make prime numbers so fascinating.

The Division Algorithm

Returning to the idea of divisors, we now examine one basic theorem that will serve as the foundation of much of modular arithmetic, which will be introduced in the next section and used in applications in the remaining sections of this chapter. This theorem is called the **division algorithm**, and it is really nothing more than the idea that when an integer is divided by a positive integer we get a quotient and a remainder. This idea is introduced in elementary school when division is first learned. We use the traditional name for this theorem, even though the division algorithm is not an algorithm. The precise statement of it is given here:

The Division Algorithm

If a is any integer and b is a positive integer, then there exist unique integers q and r such that

$$a = q \cdot b + r \quad \text{with} \quad 0 \leq r < b.$$

The number q is called the **quotient** and r is called the **remainder** in the division of a by b.

---

branching OUT

5.1

FERMAT'S LAST THEOREM: A MATH PROBLEM THAT TOOK OVER 350 YEARS TO SOLVE

Pierre de Fermat (1601–1665) was a great mathematician who pursued mathematics only as a hobby. He trained as a lawyer and worked as a magistrate. He had no formal training in higher mathematics. As was the custom at that time, he published very little of his mathematical work, and instead recorded his discoveries in letters of correspondence and in unpublished notes. Fermat is best known for a problem he suggested that turned out to be one of the most notorious mathematical problems of all time. This problem is known as Fermat's Last Theorem, and it arises naturally from the fact that there are solutions to the equation $x^2 + y^2 = z^2$, where x, y, and z are positive integers. One solution is given by the right triangle with sides of length 3, 4, and 5 (shown in the figure), where we know from the Pythagorean Theorem that $3^2 + 4^2 = 5^2$.

In fact, there are an infinite number of solutions to the equation $x^2 + y^2 = z^2$. Because of this, we might expect that equations like $x^3 + y^3 = z^3$ or $x^4 + y^4 = z^4$ also have some solutions in the positive integers. However, if you try to find such solutions, you are doomed to failure, for there are none! This is what Fermat asserted: For any positive integer $n \geq 3$, the equation $x^n + y^n = z^n$, has no solutions in the positive integers. Fermat made this claim in the margin of his copy of Diophantus' *Arithmetica*, presumably in about the year 1637. However, instead of proving this claim, he wrote in the margin, "I have assuredly found an admirable proof of this, but the margin is too narrow to contain it." Fermat's claim came to be known as Fermat's Last Theorem, though he had not given a proof of it.

Fermat's Last Theorem has a long and colorful history. Fermat himself provided a proof in the case where $n = 4$ by showing that the equation $x^4 + y^4 = z^4$ has no solutions in the positive integers, but progress on the problem beyond that point was very slow. In the eighteenth century, the great mathematician

Image © Balefire, 2012. Shutterstock, Inc.

3   5

4

Pierre de Fermat (1601–1665)

Euler provided a substantially correct proof for the case $n = 3$. The cases of $n = 5$, $n = 14$, and $n = 7$ were proven in the early nineteenth century. In 1847, Ernst Kummer showed that Fermat's Last Theorem was true for an entire class of exponents, but he still had not shown that it was true in general. In 1908, 100,000 marks was bequeathed by Paul Wolfskehl to the Academy of Science in Göttingen, to be awarded to anyone who could give a complete proof of Fermat's Last Theorem. This resulted in an onslaught of incorrect attempts by amateur mathematicians. Because only printed solutions were eligible for the prize, the problem probably has the distinction of being the mathematical problem for which the largest number of false proofs have been published. After runaway inflation in Germany in the 1920s nearly wiped out the Wolfskehl prize, it rebounded to a value of about $50,000, when it was awarded to Andrew Wiles.

Even without the draw of a large monetary reward, professional mathematicians continued to attack the now very famous problem. The armor of Fermat's Last Theorem was gradually chipped away by advances in mathematics, until finally on June 23, 1993, Professor Andrew Wiles of Princeton University announced that he had found a proof. News of his proof made the front page of the *New York Times* and a full-page article in *Time*, and it was even covered in the national evening newscasts—quite unusual for a mathematical result. *People* magazine declared Professor Wiles one of "The 25 Most Intriguing People of the Year" in 1993, putting him on the eclectic list that included, among others, Yasir Arafat, Oprah Winfrey, Howard Stern, and Michael Jordan. Professor Wiles' proof turned out to have a flaw, but after over a year of effort he corrected the proof on September 19, 1994, and it was published in 1995.

Dr. Wiles first came across Fermat's Last Theorem at the age of 10 in a book. When he finally solved the problem at age 40, Dr. Wiles confessed his bittersweet feelings about the toppling of one of mathematics' great unsolved problems. "For many of us, his problem drew us in and we always considered it something you dream about but never actually do." Now, he said, "There is a sense of loss, actually."

The question of whether or not Fermat had a proof of his theorem as he had claimed in the marginal note will never be answered. However, given the very complex nature of the mathematical ideas that Wiles used in constructing his proof, few believe that Fermat could have had a short, simple proof that somehow was not rediscovered despite serious efforts by great mathematicians over a period of 350 years.

EXAMPLE 4 *The Division Algorithm*

Find the quotient and remainder when the following divisions are performed according to the division algorithm:

(a) 51 divided by 8,
(b) 117 divided by 13.

(a) First we compute 51/8 = 6.375, and rounding this down to the nearest integer we see that $q = 6$. To find the remainder r, we write the division algorithm formula $a = q \cdot b + r$ with $a = 51$, $b = 8$, and $q = 6$ and solve for r to get

$$51 = 6 \cdot 8 + r$$
$$51 = 48 + r$$
$$r = 51 - 48 = 3.$$

Therefore, $51 = 6 \cdot 8 + 3$, and we have $q = 6$ and $r = 3$. Alternatively, we could find the quotient and remainder by performing long division.

(b) Dividing 117 by 13, we have 117/13 = 9. In this case 13 divides 117 evenly and the remainder is 0. We have $117 = 9 \cdot 13 + 0$, so $q = 9$ and $r = 0$. ∎

Notice that in the division algorithm, the number a may be a negative number. In this case we can still find a quotient q, which will be negative, such that the remainder r satisfies $0 \leq r < b$. This is illustrated in the next example.

EXAMPLE 5 *The Division Algorithm*

Find the quotient and remainder when the following divisions are performed according to the division algorithm:

(a) –57 divided by 4,
(b) –1456 divided by 16.

(a) First we compute –57/4 = –14.25. In order for the remainder r to be positive, the quotient q must be found by rounding –14.25 down to the negative integer just below it in value. So we have $q = -15$ in this case. To find r, we write out the division algorithm formula $a = q \cdot b + r$ with $a = -57$, $b = 4$, and $q = -15$ and solve for r to get

$$-57 = (-15) \cdot 4 + r$$
$$-57 = -60 + r$$
$$r = -57 + 60 = 3.$$

Therefore, $-57 = (-15) \cdot 4 + 3$, and we have $q = -15$ and $r = 3$.

(b) Because –1456/16 = –91 exactly, no rounding down is needed, and we have $-1456 = (-91) \cdot 16 + 0$, so $q = -91$ and $r = 0$. ∎

The Greatest Common Divisor

Another idea of much importance in arithmetic and number theory is that of the greatest common divisor of two numbers, which is defined as follows:

If *a* and *b* are integers with at least one of them different from 0, then the **greatest common divisor** of *a* and *b* is the largest positive integer that divides both *a* and *b*. It is denoted by

gcd(*a*, *b*).

We first run into the idea of greatest common divisor when we learn to reduce fractions in elementary school. To reduce the fraction *a/b*, we divide both the numerator and denominator by gcd(*a*, *b*). For instance, to reduce the fraction 4/6, we divide both the numerator and the denominator by gcd(4, 6) = 2 to get the reduced fraction 2/3. The idea of the greatest common divisor is important in some of the applications of number theory we will look at later in this chapter.

EXAMPLE 6 *The Greatest Common Divisor*

Find the following:

(a) gcd(8, 12),
(b) gcd(15, 0),
(c) gcd(–21, 35).

SOLUTION:

(a) The positive divisors of 8 are 1, 2, 4, and 8, and the positive divisors of 12 are 1, 2, 3, 4, 6, and 12. Therefore,

$$\gcd(8, 12) = 4.$$

(b) Because 0 is divisible by any integer, we know that it is divisible by every divisor of 15. The largest divisor of 15 is itself, and therefore

$$\gcd(15, 0) = 15.$$

(c) The positive divisors of –21 are 1, 3, 7, and 21, and the positive divisors of 35 are 1, 5, 7, and 35. Therefore,

$$\gcd(-21, 35) = 7. \qquad \blacksquare$$

TECHNOLOGY *tip*

Mathematical software, spreadsheets, and even some calculators may have the greatest common divisor function built in.

A CLOSER *look*

THE GREATEST COMMON DIVISOR AND THE EUCLIDEAN ALGORITHM

Even using the latest techniques in mathematics, it is impossible to factor most extremely large numbers quickly. It is for this reason that the RSA cryptosystem (discussed later in this chapter) is secure. You might expect that finding the greatest common divisor of two large numbers would be equally difficult. However, a method known as the *Euclidean algorithm* allows us to find the greatest common divisor of two large numbers fairly rapidly by avoiding factorization. When we apply the division algorithm to divide a by b, we obtain quotient q and remainder r. The key to the Euclidean algorithm is that the greatest common divisor of a and b is exactly the same as the greatest common divisor of b and r. We repeat the division algorithm, replacing a and b with b and r until the remainder is 0. At this stage, the number we have just divided by is the greatest common divisor of the original a and b.

Let's use the method to find the greatest common divisor of 25,443 and 80,441. We let $a = 80{,}441$, the larger of the two numbers, and $b = 25{,}443$. Then

$$80{,}441 = 3 \cdot 25{,}443 + 4112.$$

Now we apply the division algorithm with $a = 25{,}443$ and $b = 4112$, obtaining

$$25{,}443 = 6 \cdot 4112 + 771.$$

We need two more applications to get remainder 0:

$$4112 = 5 \cdot 771 + 257,$$
$$771 = 3 \cdot 257 + 0.$$

Thus, the greatest common divisor of 25,443 and 80,441 is 257.

This method allows us to compute the greatest common divisor of two 100-digit numbers in at most 478 steps. A slight refinement of the method that permits negative remainders cuts this to at most 262 steps.

Finding the greatest common divisor by listing the positive divisors of the two numbers and choosing the largest divisor they have in common, as we did in Example 6, is not a very efficient method when working with larger numbers. Another way to find $\gcd(a, b)$, where a and b are both integers greater than 1, is to first find the prime factorizations of a and b. Then $\gcd(a, b)$ is the product of all of the prime power factors common to both a and b, unless a and b have no factors in common, in which case $\gcd(a, b) = 1$. Furthermore, because

$$\gcd(a, b) = \gcd(-a, b) = \gcd(a, -b) = \gcd(-a, -b),$$

this method can also be used to find greatest common divisors involving negative numbers. This idea is illustrated in the next example.

EXAMPLE 7 *The Greatest Common Divisor*

Find the following:

(a)   gcd(660, 1400),
(b)   gcd(272, –825).

SOLUTION:

(a)   The prime factorizations of 660 and 1400 are given by

$$660 = 2^2 \cdot 3 \cdot 5 \cdot 11,$$
$$1400 = 2^3 \cdot 5^2 \cdot 7.$$

The highest power of 2 that appears in both of these prime factorizations is 2^2 and the highest power of 5 that appears in both is $5^1 = 5$. The numbers 660 and 1400 have no other prime factors in common. Therefore,

$$\gcd(660, 1400) = 2^2 \cdot 5 = 20.$$

(b)   The prime factorizations of 272 and 825 are

$$272 = 2^4 \cdot 17,$$
$$825 = 3 \cdot 5^2 \cdot 11.$$

Because these prime factorizations have no factors in common, we see that

$$\gcd(272, 825) = 1,$$

and therefore

$$\gcd(272, -825) = 1. \qquad \blacksquare$$

Two integers a and b with gcd(a, b) = 1, like 272 and –825 in Example 7, are called **relatively prime**.

Now that we have a basic foundation in the ideas of primes and divisibility, we will go on to use these ideas in the next section to develop a new kind of arithmetic, called modular arithmetic, in which divisibility plays a key role.

EXERCISES / FOR SECTION 5.1

In Exercises 1–12, determine whether the statement is true or false.

1. $7|42$
2. $4|18$
3. $12|40$
4. $18|10$
5. $0|28$
6. $32|160$
7. $6|-34$
8. $15|-55$
9. $4|0$
10. $0|24$
11. $-10|240$
12. $-5|20$

In Exercises 13–18, find the prime factorization of the given number.

13. 132
14. 112
15. 1625
16. 1755
17. 6800
18. 8100

19. Use the Sieve of Eratosthenes to find all primes less than or equal to 100.
20. Use the Sieve of Eratosthenes to find all primes less than or equal to 150.

In Exercises 21–26, determine whether or not the given number is prime.

21. 751
22. 979
23. 1817
24. 1247
25. 2579
26. 2131

27. Show that the Fermat number $F_3 = 2^{2^3} + 1$ is prime.
28. Show that $n^2 - n + 41$ is prime for $n = 22$.

Goldbach's conjecture states that every even integer greater than 4 can be written as a sum of two odd prime numbers. In Exercises 29–34, write the given number as the sum of two odd prime numbers. (There may be more than one possible answer.)

29. 16
30. 18
31. 46
32. 38
33. 98
34. 54

In 1949, H. E. Richert proved that every integer $n > 6$ can be written as a sum of one or more distinct primes. In Exercises 35–38, write the given number in this way. (There may be more than one possible answer.)

35. 86
36. 73
37. 232
38. 147

In Exercises 39–48, find the quotient and remainder when the division is performed according to the division algorithm.

39. 68 divided by 7
40. 92 divided by 8
41. 204 divided by 12
42. 192 divided by 11
43. −36 divided by 5
44. −18 divided by 4
45. −123 divided by 9
46. −270 divided by 15
47. −87 divided by 10
48. −55 divided by 6

In Exercises 49–60, evaluate the greatest common divisor.

49. gcd(9, 15)
50. gcd(18, 30)
51. gcd(0, 24)
52. gcd(–25, –12)
53. gcd(–16, 40)
54. gcd(6, 0)
55. gcd(77, 130)
56. gcd(168, 700)
57. gcd(650, 475)
58. gcd(170, –81)
59. gcd(–525, –231)
60. gcd(700, 900)

For Exercises 61–63, consider the following: A positive integer is called a *perfect number* if it is equal to the sum of all of its positive divisors except itself. For instance, the number 6 is perfect because 6 = 1 + 2 + 3. It is not known whether there are infinitely many perfect numbers or whether there are any odd perfect numbers.

61. Show that the number 10 is not perfect.
62. Find the next perfect number after 6.
63. The third perfect number is 496. Verify that it is perfect.
64. Suppose a positive integer n has at least three prime factors. Let p be the smallest prime factor of n. In terms of n, what is the largest that p can be? Explain your answer.
65. (a) For what positive integers n is $n(n + 1)/2$ an integer? Explain your answer.
    (b) For what positive integers n is $n(n + 1)/2$ an even integer? Explain your answer.
66. Find all positive integers n for which $n^2 - 1$ is a prime number. Explain how you know you have found them all.

5.2 MODULAR ARITHMETIC

Many applications of number theory use a special kind of arithmetic known as **modular arithmetic**. In modular arithmetic, rather than thinking of integers in the usual way, we instead consider only their remainders after division by a given integer n. In this way, modular arithmetic can be thought of as an arithmetic of remainders. The idea of modular arithmetic was first introduced by the German mathematician Carl Friedrich Gauss (1777–1855) in his work *Disquisitiones Arithmeticae*. Gauss is considered by many to be the greatest mathematician who ever lived, and *Disquisitiones Arithmeticae* was published when he was only 24 years old.

It may at first seem strange to be concerned only with the remainder of an integer after division by a particular integer, but such situations arise fairly naturally. For instance, if a clock now reads 7:00, and you want to know what time the clock will read 38 hours from now, you need only know the remainder when 7 + 38 = 45 is divided by 12. By the division algorithm, 45 = 3 · 12 + 9. Therefore, because 9 is the remainder, the clock will read 9:00. In the remaining sections of this chapter, we

branching OUT

CARL FRIEDRICH GAUSS: THE PRINCE OF MATHEMATICIANS

It is difficult, if not impossible, to bestow the title of the "best mathematician who ever lived" to any individual. Because the work that mathematicians do depends so much on the previous work of mathematicians, it is impossible to judge just who made the most remarkable discoveries. Nevertheless, many argue that Carl Friedrich Gauss is the greatest mathematician of all time.

Unlike most other distinguished scholars of his time, Gauss did not come from a privileged

Carl Friedrich Gauss (1777–1855).

Image © Nicku, 2012. Shutterstock, Inc.

family. He was born in Germany in 1777 to a poor and uneducated family, but as a young child his intellectual gifts were apparent. His mathematical prowess was so great that when Gauss reached the age of 10, his schoolmasters claimed that they had nothing else to teach him. Fortunately, his gifts caught the attention of Duke Ferdinand of Brunswick, who saw to it that Gauss received a proper education. While studying at the University of Göttingen and when he was only 18 years old, Gauss proved that a regular 17-gon could be constructed with only an unmarked ruler and compass. This was an amazing discovery, for at the time no new constructions of regular polygons had been discovered since the time of Euclid and no one knew that such a construction was even possible. This discovery was just the first of many in Gauss' long and distinguished career. Although Gauss made major contributions to many areas of mathematics, he was particularly fond of number theory. He is credited with saying "Mathematics is the Queen of the sciences, and the theory of numbers is the Queen of mathematics." His publication, *Disquisitiones Arithmeticae*, laid the foundation for the field of modern number theory.

will see many practical applications that are made easier through the use of modular arithmetic.

In modular arithmetic, we have the special notation for remainders given next.

If n is a positive integer and a is any integer, then

$$a \bmod n$$

is the remainder obtained when a is divided by n according to the division algorithm. Note that by the definition of the remainder, we know that $0 \le a \bmod n < n$.

EXAMPLE 1 *Computing a mod n*

Evaluate each of the following quantities:

(a) 11 mod 4,
(b) 253 mod 35,
(c) 18 mod 6,
(d) −43 mod 5,
(e) −1272 mod 24.

SOLUTION:

(a) By the division algorithm, $11 = 2 \cdot 4 + 3$, so the remainder of 11 when divided by 4 is 3. Therefore, 11 mod 4 = 3.
(b) Because $253 = 7 \cdot 35 + 8$, we have 253 mod 35 = 8.
(c) We have $18 = 3 \cdot 6 + 0$, so 18 mod 6 = 0.
(d) We have $-43 = (-9) \cdot 5 + 2$, and therefore −43 mod 5 = 2.
(e) Because $-1272 = (-53) \cdot 24 + 0$, it follows that −1272 mod 24 = 0.

Note that parts (c) and (e) of Example 1 illustrate the fact that $n \mid a$ is equivalent to a mod $n = 0$. ∎

TECHNOLOGY *tip*

Most number theory or general mathematics software, spreadsheets, and some calculators can compute a mod n automatically.

In modular arithmetic, we work with one fixed **modulus** n and call the arithmetic we are doing **arithmetic modulo n**. To define what we mean by arithmetic modulo n, first we must define what it means for two numbers to be "equal" modulo n. Rather than using the word equal to signify that two things are equivalent in modular arithmetic, we use the word congruent. The precise definition for two numbers to be congruent modulo n is as follows:

> **Congruence Modulo n**
>
> Let n be a positive integer. Two numbers a and b are said to be **congruent modulo n** if they both have the same remainder when divided by n—in other words, if
>
> $$a \bmod n = b \bmod n.$$
>
> We use the notation
>
> $$a \equiv b \ (\mathrm{mod}\ n)$$
>
> to indicate that a is congruent to b modulo n. If a is not congruent to b modulo n, we write
>
> $$a \not\equiv b \ (\mathrm{mod}\ n).$$

A commonly used equivalent definition is that a and b are congruent modulo n if $n \mid (a - b)$.

EXAMPLE 2 *Congruence Modulo n*

(a) Determine whether 35 is congruent to 67 modulo 8.
(b) Determine whether –26 is congruent to 15 modulo 12.

SOLUTION:

(a) Because 35 leaves remainder 3 when divided by 8 and 67 leaves remainder 3 when divided by 8, we see that 35 mod 8 = 67 mod 8 = 3, and therefore $35 \equiv 67$ (mod 8).

(b) By the division algorithm, $-26 = (-3) \cdot 12 + 10$, and therefore –26 mod 12 = 10. Because 15 mod 12 = 3, it follows that $-26 \not\equiv 15$ (mod 12). ∎

We are now ready to introduce arithmetic modulo n. The basic rules of addition, subtraction, and multiplication are similar to those we use in regular arithmetic. The main difference is that two numbers or calculations are equal if they are congruent modulo n. Three major properties of modular arithmetic are listed next.

Properties of Modular Arithmetic

If $a \equiv b$ (mod n) and $c \equiv d$ (mod n), then

1. $a + c \equiv b + d$ (mod n),
2. $a - c \equiv b - d$ (mod n),
3. $ac \equiv bd$ (mod n).

For example, because $14 \equiv 4$ (mod 5) and $48 \equiv 3$ (mod 5), Property 1 tells us that

$$14 + 48 \equiv 4 + 3 \text{ (mod 5)}.$$

Because $14 + 48 = 62$ and $4 + 3 = 7$, we see that

$$62 \equiv 7 \text{ (mod 5)}.$$

Similarly, Property 2 tells us that

$$14 - 48 \equiv 4 - 3 \text{ (mod 5)},$$

or

$$-34 \equiv 1 \text{ (mod 5)}.$$

It follows from Property 3 that

$$14 \cdot 48 \equiv 4 \cdot 3 \ (\text{mod } 5),$$

or

$$672 \equiv 12 \ (\text{mod } 5).$$

When we are working in arithmetic modulo n, we usually want the numbers we are working with to be small. The properties of modular arithmetic allow us to substitute a small number b in place of a larger number a in calculations involving addition, subtraction, and multiplication modulo n as long as $a \equiv b \ (\text{mod } n)$. In applications, we usually want our final answer to lie between 0 and $n - 1$, so we want to give the answer mod n. These ideas are illustrated in the following example.

EXAMPLE 3   *Modular Arithmetic*

Perform the following calculations. In each case, the final answer must be a number between 0 and $n - 1$, where n is the modulus:

(a) $(17 + 86) \bmod 4,$
(b) $(23 \cdot 32 \cdot 2) \bmod 20,$
(c) $(157(13 - 16)) \bmod 17,$
(d) $(37^5 - 3 + 42) \bmod 7.$

SOLUTION:

(a)   Using the fact that $17 \equiv 1 \ (\text{mod } 4)$ and $86 \equiv 2 \ (\text{mod } 4)$, we see that

$$17 + 86 \equiv 1 + 2 \equiv 3 \ (\text{mod } 4).$$

Because $0 \le 3 < 4$, we have

$$(17 + 86) \bmod 4 = 3.$$

(b)   In this case we use the fact that $23 \equiv 3 \ (\text{mod } 20)$ and $32 \equiv 12 \ (\text{mod } 20)$ to find that

$$23 \cdot 32 \cdot 2 \equiv 3 \cdot 12 \cdot 2 \equiv 72 \equiv 12 \ (\text{mod } 20),$$

and therefore, because $0 \le 12 < 20$,

$$(23 \cdot 32 \cdot 2) \bmod 20 = 12.$$

(c)   Using the fact that $157 \equiv 4 \ (\text{mod } 17)$, we have

$$157(13 - 16) \equiv 4(-3) \equiv -12 \equiv 5 \ (\text{mod } 17),$$

so, noting that $0 \le 5 < 17$,

$$(157(13 - 16)) \bmod 17 = 5.$$

(d)   We know $37 \equiv 2 \pmod 7$ and $42 \equiv 0 \pmod 7$, so

$$37^5 - 3 + 42 \equiv 2^5 - 3 + 0 \equiv 32 - 3 \equiv 29 \equiv 1 \pmod 7.$$

Because $0 \le 1 < 7$, we have

$$(37^5 - 3 + 42) \bmod 7 = 1. \qquad ■$$

There are instances where it is helpful to substitute a negative number for a positive number in a modular arithmetic calculation, as in the following example.

EXAMPLE 4   *Modular Arithmetic*

Find $32^{50} \bmod 33$.

SOLUTION:

The number 32^{50} is quite large. However, if we use the fact that $32 \equiv -1 \pmod{33}$, we have

$$32^{50} \equiv (-1)^{50} \equiv ((-1)^2)^{25} \equiv 1^{25} \equiv 1 \pmod{33}.$$

Therefore, because $0 \le 1 < 33$,

$$32^{50} \bmod 33 = 1. \qquad ■$$

The properties of modular arithmetic provide a simple way to find a negative number mod n, as illustrated in the following example.

EXAMPLE 5   *Finding a Negative Number mod n*

Evaluate $-136 \bmod 7$.

SOLUTION:

Because $7 \equiv 0 \pmod 7$, adding multiples of 7 does not change the value of a quantity in arithmetic modulo 7. To evaluate $-136 \bmod 7$, we add just enough multiples of 7 to change the value -136 to a number between 0 and 6 in arithmetic modulo 7. In this case, because $136/7 \approx 19.43$, we add $20 \cdot 7$ to get

$$-136 \equiv -136 + 20 \cdot 7 \equiv -136 + 140 \equiv 4 \pmod 7,$$

so

$$-136 \bmod 7 \equiv 4. \qquad \blacksquare$$

You may have noticed that we have added, subtracted, and multiplied in modular arithmetic, but we have not divided. In regular arithmetic, we can divide by any number except 0. In arithmetic modulo n, it is only possible to divide by a number k for which $\gcd(k, n) = 1$. We will address this idea later in this chapter when it comes up in some of our applications.

EXERCISES FOR SECTION 5.2

In Exercises 1–20, evaluate the given quantity.

1. 48 mod 9
2. 8 mod 3
3. −38 mod 12
4. −64 mod 8
5. 396 mod 7
6. 67 mod 8
7. 14 mod 2
8. −2115 mod 4
9. −342 mod 6
10. 135 mod 24
11. −54 mod 4
12. −73 mod 10
13. 45,237 mod 10
14. −136 mod 7
15. −252 mod 11
16. 461 mod 12
17. 2295 mod 16
18. −3 mod 9
19. −30,881 mod 50
20. 4312 mod 5

In Exercises 21–40, perform the given calculation. In each case, the answer must be a number between 0 and $n - 1$, where n is the modulus.

21. $(9 + 14) \bmod 5$
22. $(46 + 27) \bmod 6$
23. $(33 \cdot 96 \cdot 11) \bmod 10$
24. $(32 - 58) \bmod 2$
25. $(15^3 - 25) \bmod 4$
26. $(44 \cdot 38 \cdot 9 \cdot 72) \bmod 7$
27. $(47(42 - 18)) \bmod 9$
28. $12^{99} \bmod 3$
29. $(8 \cdot 9 \cdot 10 \cdot 11 \cdot 12 \cdot 13) \bmod 7$
30. $(42(32 - 45)) \bmod 12$
31. $(2^{310} + 1) \bmod 3$
32. $(14^{387} \cdot 16^{563}) \bmod 15$
33. $(83 + 30 + 15 + 3) \bmod 9$
34. $(31 \cdot 34) \bmod 30$
35. $(47 \cdot 50) \bmod 44$
36. $((15 + 18)(61 - 90)) \bmod 2$
37. $(42^5 + 17^{28}) \bmod 8$
38. $(456 + 321 + 9872 + 400) \bmod 10$
39. $(823 + 620) \bmod 100$
40. $(23 \cdot 38) \bmod 7$

If a positive integer is divided by 10, then the remainder is its last digit. Therefore, a positive integer is congruent to its last digit mod 10. In Exercises 41–44, use this fact and modular arithmetic to find the last digit of the given number.

41. 9^{2001}
42. 3^{1000}
43. 7453^{56}
44. $123{,}456{,}789^{999}$

45. To pass the time, a general lines up his soldiers in 100 equal rows, with three soldiers left over. He then lines them up in 99 equal rows, this time with two soldiers left over. The general has under 10,000 soldiers. Exactly how many soldiers does he have? Explain how you arrived at your answer.

46. Two pirates were marooned on an island with a monkey and a "treasure" consisting of a pile of coconuts. Not trusting the other pirate, the first pirate rises in the night, divides the coconuts evenly into two piles, with one coconut left over, which he gives to the monkey. He then hides his pile of coconuts, leaves the remaining coconuts, and goes back to sleep. Later, the second pirate rises, divides the remaining coconuts evenly into two piles, with one coconut left over, which he also gives to the monkey. After hiding his pile of coconuts, he leaves the remaining coconuts and then goes back to sleep. Morning comes. The pirates divide their "treasure" into halves, with one coconut left over for the monkey.

(a) Find the number of coconuts mod 8 in the original pile.

(b) What is the smallest number of coconuts the original pile could have contained?

5.3 DIVISIBILITY TESTS

A **divisibility test** is a method for determining whether a given integer is divisible by another given integer without actually performing the division. There are several simple divisibility tests that make use of modular arithmetic. These tests can be used not only to test divisibility, but, as we will see later, also to check calculations.

The divisibility tests we will look at make use of the fact that our numbers are written in base 10 notation. For example, the number 357 stands for

$$357 = 3 \cdot 100 + 5 \cdot 10 + 7 = 3 \cdot 10^2 + 5 \cdot 10 + 7,$$

and the expression $3 \cdot 10^2 + 5 \cdot 10 + 7$ is called the base 10 expansion of 357. Similarly, the base 10 expansion of the number 2,508,623 is given by

$$2,508,623 = 2 \cdot 10^6 + 5 \cdot 10^5 + 0 \cdot 10^4 + 8 \cdot 10^3 + 6 \cdot 10^2 + 2 \cdot 10 + 3.$$

In general, the base 10 expansion of a positive integer $N = a_m a_{m-1} \cdots a_1 a_0$, where the number a_k is the $(k + 1)$st digit of N from the right, is given by

$$N = a_m \cdot 10^m + a_{m-1} \cdot 10^{m-1} + \cdots + a_1 \cdot 10 + a_0.$$

The first divisibility test we will look at is one for testing divisibility by 9. This test is based on the fact that

$$10 \equiv 1 \ (\text{mod } 9).$$

Therefore, in arithmetic modulo 9, we can substitute 1 for 10 in the base 10 expansion for $N = a_m a_{m-1} \cdots a_1 a_0$ to get

$$N \equiv a_m \cdot 10^m + a_{m-1} \cdot 10^{m-1} + \cdots + a_1 \cdot 10 + a_0$$
$$\equiv a_m \cdot 1^m + a_{m-1} \cdot 1^{m-1} + \cdots + a_1 \cdot 1 + a_0$$
$$\equiv a_m + a_{m-1} + \cdots + a_1 + a_0 \ (\text{mod } 9).$$

In other words, N is congruent to the sum of its digits modulo 9. Because N mod 9 is the remainder we obtain when dividing N by 9, it follows that $(a_m + a_{m-1} + \cdots + a_1 + a_0)$ mod 9 is the remainder we obtain when dividing N by 9. When adding the digits modulo 9, we can ignore any 9's or any combination of digits that add up to 9 because they are congruent to 0 modulo 9. This process for finding the remainder obtained after dividing by 9 is called **casting out nines**. Because a number is divisible by 9 if and only if the remainder obtained after dividing by 9 is 0, we have the following divisibility test:

A Test for Divisibility by 9 by Casting Out Nines

A positive integer is divisible by 9 if and only if the sum of its digits mod 9 is 0.

We illustrate this divisibility test in the following example.

EXAMPLE 1 *Casting Out Nines*

Use casting out nines to determine whether or not the following numbers are divisible by 9:

(a) 37,940,531,
(b) 6,206,309,719,290.

SOLUTION:

(a) Summing the digits of 37,940,531 modulo 9, casting out nines, and finding combinations of digits that add up to 9, we get

$$3 + 7 + 9 + 4 + 0 + 5 + 3 + 1 \equiv 3 + 7 + 4 + (5 + 3 + 1) \ (\text{mod } 9).$$

Now casting out the combination of digits that adds up to 9, we have

$$3 + 7 + 9 + 4 + 0 + 5 + 3 + 1 \equiv 3 + 7 + 4 \equiv 14 \equiv 5 \ (\text{mod } 9).$$

Therefore, the sum of the digits mod 9 is 5, so 37,940,531 is not divisible by 9 (and, in fact, it leaves remainder 5 when divided by 9).

(b) Adding up the digits of 6,206,309,719,290 modulo 9 and casting out nines and combinations of digits that add up to 9, we have

$$6 + 2 + 0 + 6 + 3 + 0 + 9 + 7 + 1 + 9 + 2 + 9 + 0$$

$$\equiv 6 + 2 + (6 + 3) + 1 + (7 + 2)$$

$$\equiv (6 + 2 + 1) \equiv 0 \ (\text{mod } 9).$$

Therefore, 6,206,309,719,290 is divisible by 9. ∎

Using the same ideas as we did in arriving at casting out nines, we can find a test for divisibility by 11. We start with the fact that

$$10 \equiv -1 \ (\text{mod } 11).$$

If $N = a_m a_{m-1} \cdots a_1 a_0$, then in arithmetic modulo 11 we can substitute -1 for 10 in the decimal expansion of N to get

$$N \equiv a_m \cdot 10^m + a_{m-1} \cdot 10^{m-1} + \cdots + a_1 \cdot 10 + a_0$$

$$\equiv a_m \cdot (-1)^m + a_{m-1} \cdot (-1)^{m-1} + \cdots + a_1 \cdot (-1) + a_0.$$

Writing the sum in reverse order, we see that

$$N \equiv a_0 - a_1 + a_2 - a_3 + \cdots + (-1)^m \cdot a_m \ (\text{mod } 11).$$

Thus N is congruent to the "alternating sum" of its digits modulo 11, where by an alternating sum of its digits we mean the first digit on the right minus the second digit from the right plus the third digit, and so on. Because $N \bmod 11$ is the remainder we obtain after dividing N by 11, it follows that $(a_0 - a_1 + a_2 - a_3 + \cdots + (-1)^m \cdot a_m) \bmod$ 11 is the remainder we obtain after dividing N by 11. When calculating the alternating sum mod 11, many of the numbers will cancel out with regular addition. We can also throw out any combination of digits that add up to 11 or -11, so this method of finding $N \bmod 11$ is sometimes called **casting out elevens**. We have the following test for divisibility by 11:

A Test for Divisibility by 11 by Casting Out Elevens

A positive integer $N = a_m a_{m-1} \cdots a_1 a_0$ is divisible by 11 if and only if the alternating sum of its digits mod 11 is 0—in other words, if and only if

$$(a_0 - a_1 + a_2 - a_3 + \cdots + (-1)^m \cdot a_m) \bmod 11 = 0.$$

EXAMPLE 2 *Casting Out Elevens*

Use casting out elevens to determine whether or not the following numbers are divisible by 11:

(a) 4,506,073,
(b) 771,003,239.

SOLUTION:

(a) The alternating sum of the digits of 4,506,073 mod 11 is given by

$$3 - 7 + 0 - 6 + 0 - 5 + 4 \equiv 3 - 7 - (6 + 5) + 4$$
$$\equiv 3 - 7 + 4 \equiv 0 \ (\text{mod } 11),$$

and therefore 4,506,073 is divisible by 11.

(b) For 771,003,239, the alternating sum of the digits mod 11 is

$$9 - 3 + 2 - 3 + 0 - 0 + 1 - 7 + 7 \equiv (9 + 2) - 3 - 3 + 1$$
$$\equiv -5 \equiv -5 + 11 \equiv 6 \ (\text{mod } 11),$$

and therefore 771,003,239 is not divisible by 11. ■

We derive one more divisibility test, that for divisibility by 4. As you might guess by now, the idea for this comes from looking at the decimal expansion of the number $N = a_m a_{m-1} \cdots a_1 a_0$ modulo 4. First we note that $4 \,|\, 10^2$, so $10^2 \equiv 0 \ (\text{mod } 4)$. In general, we can see that because $4 \,|\, 10^k$ for $k \geq 2$, we have

$$10^k \equiv 0 \ (\text{mod } 4) \quad \text{for } k \geq 2.$$

Therefore, in arithmetic modulo 4, whenever $k \geq 2$ we can substitute 0 for 10^k in the base 10 expansion for $N = a_m a_{m-1} \cdots a_1 a_0$ to get

$$N \equiv a_m \cdot 10^m + a_{m-1} \cdot 10^{m-1} + \cdots + a_2 \cdot 10^2 + a_1 \cdot 10 + a_0$$
$$\equiv a_m \cdot 0 + a_{m-1} \cdot 0 + \cdots + a_2 \cdot 0 + a_1 \cdot 10 + a_0$$
$$\equiv a_1 \cdot 10 + a_0 \ (\text{mod } 4).$$

However, $a_1 \cdot 10 + a_0$ is just the two-digit number $a_1 a_0$ given by the last two digits of N. It follows that N and the number given by its last two digits are congruent modulo 4 and therefore leave the same remainder when divided by 4. Thus, we have the following test for divisibility by 4:

Test for Divisibility by 4

A positive integer $N = a_m a_{m-1} \cdots a_1 a_0$ is divisible by 4 if and only if the two-digit number $a_1 a_0$ given by its last two digits is divisible by 4.

EXAMPLE 3 *Divisibility by 4*

Use the test for divisibility by 4 to determine whether or not the following numbers are divisible by 4:

(a) 9,903,562,748,
(b) 7,903,603.

SOLUTION:

(a) The number formed by the last two digits of 9,903,562,748 is 48, and 48 = 4 · 12 is divisible by 4. Therefore, 9,903,562,748 is also divisible by 4.
(b) The number formed by the last two digits of 7,903,603 is 03 = 3. Because 3 is not divisible by 4, we see that 7,903,603 is not divisible by 4. ■

There are tests for divisibility by other numbers that are based on ideas similar to those used in the tests for divisibility by 9, 11, and 4. A few of these tests are summarized here.

Other Divisibility Tests

A positive integer is divisible by

- 2 if and only if its last digit is even,
- 3 if and only if the sum of its digits mod 3 is 0,
- 5 if and only if its last digit is 0 or 5,
- 6 if and only if it is divisible by both 2 and 3,
- 7 if and only if when its last digit is doubled and then subtracted from the number formed by the remaining digits, the resulting number is divisible by 7,
- 8 if and only if the number formed by its last three digits is divisible by 8,
- 10 if and only if its last digit is 0.

We will look at all of these tests in the exercises at the end of this section. In the next example, we see how the test for divisibility by 7 works.

EXAMPLE 4 *Divisibility by 7*

Use the test for divisibility by 7 to determine whether or not 4183 is divisible by 7.

When the last digit, 3, of 4183 is doubled and subtracted from the remaining digits, we have

$$
\begin{array}{r}
418 \\
-6 \\
\hline
412
\end{array}
$$

Now we see that 4183 is divisible by 7 if and only if 412 is divisible by 7. To see whether 412 is divisible by 7, we apply the test for divisibility by 7 to it. Doubling the last digit of 412 and subtracting it from the remaining digits gives

$$
\begin{array}{r}
41 \\
-4 \\
\hline
37
\end{array}
$$

We know that 37 is not divisible by 7. Therefore, we can conclude that 412 is not divisible by 7, and hence, neither is the number we started with, 4183. ■

Sometimes divisibility tests can supply missing information about a number, as in the next example.

EXAMPLE 5 *Finding the Missing Digit*

Suppose you purchased 22 identically priced drinks with a check. The entry in the checkbook is difficult to read, but you know the total price was \$14.#6, where # is an illegible digit. Assuming there was no sales tax, what is the illegible digit?

Because $22 = 2 \cdot 11$ divides into the number 14#6, we know that $11 \mid 14\#6$. Therefore, the alternating sum of the digits of 14#6 must be congruent to 0 mod 11. This gives us the equation

$$
6 - \# + 4 - 1 \equiv 9 - \# \equiv 0 \ (\text{mod } 11).
$$

Therefore, the digit # must be a 9. ■

Now we will see how divisibility tests can be used to provide a partial check on an arithmetic calculation. In particular, we will look at how casting out nines can be used to check multiplication. Suppose we want to multiply two positive integers N and

M. If, for instance, N mod 9 = 5 and M mod 9 = 3, then by the properties of modular arithmetic we know that $N \cdot M \equiv 5 \cdot 3 \equiv 15 \equiv 6 \pmod 9$. Therefore, after we have multiplied N and M in regular arithmetic, we can partially check the calculation by seeing if the answer is 6 modulo 9. If it is not, then we know the calculation is incorrect. This idea is illustrated in the next example.

EXAMPLE 6 *Checking a Calculation*

Suppose you multiply the two numbers 4569 and 3428 and arrive at the answer 15,662,542. Use casting out nines to check the answer.

SOLUTION:

We know that every number is congruent to the sum of its digits modulo 9, so we have

$$4569 \equiv 4 + 5 + 6 + 9 \equiv (4 + 5) + 6 \equiv 6 \pmod 9$$

and

$$3428 \equiv 3 + 4 + 2 + 8 \equiv (3 + 4 + 2) + 8 \equiv 8 \pmod 9.$$

Therefore, we know from the properties of modular arithmetic that

$$4569 \cdot 3428 \equiv 6 \cdot 8 \equiv 48 \equiv 3 \pmod 9.$$

Now finding what the answer 15,662,542 is modulo 9 by summing its digits modulo 9, we have

$$15{,}662{,}542 \equiv 1 + 5 + 6 + 6 + 2 + 5 + 4 + 2$$
$$\equiv (1 + 6 + 2) + 5 + 6 + (5 + 4) + 2$$
$$\equiv 5 + 6 + 2 \equiv 13 \equiv 4 \pmod 9.$$

Because this does not agree with the fact that $(4569 \cdot 3428)$ mod 9 = 3, we know that the calculation $4569 \cdot 3728 = 15{,}662{,}542$ must be wrong. (In fact, the correct calculation is $4569 \cdot 3428 = 15{,}662{,}532$.) ■

It is important to note that casting out nines provides only a partial check that a calculation has been done correctly. It is quite possible to arrive at an incorrect answer that just happens to be congruent to the correct answer modulo 9. For example, suppose we multiply 29 and 84 and arrive at the incorrect answer 2976. (The correct answer is 2436.) We have $29 \equiv 2 + 9 \equiv 2 \pmod 9$ and $84 \equiv 8 + 4 \equiv 12 \equiv 3 \pmod 9$, and therefore

$$29 \cdot 84 \equiv 2 \cdot 3 \equiv 6 \pmod 9.$$

Because

$$2976 \equiv 2 + 9 + 7 + 6 \equiv (2 + 7) + 6 \equiv 6 \ (\text{mod } 9),$$

our check does not tell us that the calculation $29 \cdot 84 = 2976$ is incorrect. However, even with this flaw, casting out nines does catch most errors, so it provides a nice partial check. We can also use casting out elevens or other divisibility tests to check calculations.

branching OUT

CALCULATING A BILLION DIGITS OF PI

We know that a circle of radius r has area equal to πr^2 and perimeter equal to $2\pi r$, where the constant π is a number approximately equal to 3.14. But what is the exact value of π? The Greek scholar Archimedes, in his work *On the Measurement of the Circle*, approximated π by inscribing and circumscribing polygons in a circle as in the figure and computing the perimeters of those polygons.

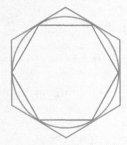

Using this method with regular polygons with 96 sides, Archimedes showed that $3\frac{10}{71} < \pi < 3\frac{1}{7}$ or in decimal notation $3.140845\ldots < \pi < 3.142857\ldots$.

In 1771, Johann Lambert (1728–1777) proved that π is irrational, and therefore that its decimal expansion is not finite nor does it repeat in a pattern. Before Lambert's proof, π had been calculated to 112 decimal places:

$\pi \approx 3.1415926535897932384626433832795028841$
$9716939937510582097494459230781640628$
$62089986280348253421170679821480865513.$

The search for even more precise calculations of π has been carried out by both amateur and professional mathematicians as a kind of sport.

In 1989, brothers Gregory V. and David V. Chudnovsky of Columbia University calculated over a billion digits of π. Using the formula

$$\pi = \left(\frac{1}{426{,}880\sqrt{10{,}005}} \sum_{n=0}^{\infty} \left[\frac{(6n)!(545{,}140{,}134n + 13{,}591{,}409)}{n!^3(3n)!(-640{,}320)^{3n}} \right] \right)^{-1}$$

and the help of high-powered computers, they computed π to 1,011,196,691 digits. You might think it is unlikely for a computer to make a calculation error if it has been programmed correctly, but in the context of calculations of this magnitude, errors are to be expected. Because of the possibility of calculation errors on the computer, the Chudnovsky brothers used modular arithmetic to check the calculations as they went along. Using these modular arithmetic checks along with other checks, the probability of an undetected error at any step was less than 1 in 10^{290}.

As of early 2012, the record for the number of digits of π was just over 10 trillion digits, calculated by Alexander Yee and Shigeru Kondo. This staggering number of digits would fill about 200 million pages.

EXERCISES FOR SECTION 5.3

In Exercises 1–6, use casting out nines to determine whether or not the given number is divisible by 9.

1. 7893
2. 327
3. 1,957,195,819,861,989
4. 123,456,789,123,456,789
5. 5,356,990,218,031,912,008
6. 80,907,704,561,223,513,069

In Exercises 7–12, use casting out elevens to determine whether or not the given number is divisible by 11.

7. 9123
8. 51,457
9. 741,050,902
10. 45,477,432
11. 123,456,789,987,654,321
12. 43,991,704,651,030,209

In Exercises 13–16, use the test for divisibility by 4 to determine whether or not the given number is divisible by 4.

13. 78,406
14. 816
15. 82,168,003,234,084
16. 780,811,234

In Exercises 17–20, use the divisibility tests to determine whether or not the given number is divisible by each of the following:

(a) 2   (b) 3   (c) 4   (d) 5   (e) 6
(f) 7   (g) 8   (h) 9   (i) 10   (j) 11

In each case, show how you used the divisibility test to arrive at your answer.

17. 62,952
18. 857,304
19. 2,756,768,175
20. 531,468,300

In Exercises 21–24, use arguments similar to the ones used in the text to arrive at the tests for divisibility by 9, 11, and 4.

21. Show that a positive integer is divisible by 8 if and only if the number formed by its last three digits is divisible by 8.
22. Show that a positive integer is divisible by 5 if and only if its last digit is 0 or 5.
23. Show that a positive integer is divisible by 3 if and only if the sum of its digits mod 3 is 0.
24. Come up with a test for divisibility by 16 (it is similar to those for divisibility by 4 and 8) and show why the test works.
25. A child bought nine identical candy bars for a total of \$4.#1, where the missing digit was smeared with chocolate. Assuming there was no sales tax, how much did each bar cost?
26. A storekeeper purchased 144 identically priced hammers from a wholesaler. When setting her selling price for the hammers, she checked the shop's record for the wholesale price. Her accountant had written

$1#25.4#,

where the #'s represent illegible digits, which may or may not be different. Determine the wholesale price of a single hammer.

27. A *palindrome* is a number that reads the same backwards as forwards. For example, the numbers 151 and 12,033,021 are palindromes. Explain why any palindrome with an even number of digits is divisible by 11.

28. Use casting out nines to determine whether the following calculation *could* be correct:

$$89,034,690,257 \cdot 739,005,139,645 = 65,797,093,306,623,605,938,765.$$

29. Use casting out nines to determine whether the following calculation *could* be correct:

$$730,963,709 \cdot 28,100,923,561,398 = 20,450,755,312,764,971,305,182.$$

30. Use casting out elevens to determine whether the calculation in Exercise 28 *could* be correct.

31. Use casting out elevens to determine whether the calculation in Exercise 29 *could* be correct.

5.4   CHECK DIGITS

Identification numbers are widespread in society today. We find them on credit cards, driver's licenses, retail items, bank checks, and many other places. For many of these kinds of numbers, the last digit is an extra digit that is mathematically related to all of the other digits in the number. We call this type of special digit a **check digit**. As we will see, the check digit provides a way to detect errors that may occur when the number is recorded. The check digit also makes it somewhat more difficult to come up with fraudulent identification numbers for items like credit cards and bank checks because the person making up the identification numbers would have to know what method is normally used to generate the check digit. There are many different methods used to generate check digits, and modular arithmetic plays a role in most of these methods. We begin by looking at some of the simpler methods, and then go on to look at some more complicated methods.

Throughout our discussion of identification numbers, we will refer to the part of an identification number that does not include the check digit as the **main part** of the number.

The check digit method used for the serial numbers on U.S. Postal Service money orders is fairly simple. The serial numbers consist of eleven digits, the first ten of which are the main part, and the eleventh digit is a check digit that is equal to the main part mod 9. Suppose a money order has serial number 19187790107. Recalling that a number is congruent to the sum of its digits modulo 9, we have

$$1918779010 \equiv 1 + 9 + 1 + 8 + 7 + 7 + 9 + 0 + 1 + 0$$
$$\equiv 1 + (1 + 8) + 7 + 7 + 1 \equiv 16 \equiv 7 \ (\text{mod } 9),$$

so the check digit 7 is equal to the main part of the serial number mod 9 as we would expect.

EXAMPLE 1 *U.S. Postal Service Money Order*

Suppose a document that appears to be a U.S. Postal Service money order has serial number 56347894283. Could it be an authentic money order?

SOLUTION:

The check digit for this serial number should be

$$5634789428 \equiv 5 + 6 + 3 + 4 + 7 + 8 + 9 + 4 + 2 + 8$$
$$\equiv (5 + 4) + (6 + 3) + (7 + 2) + 8 + 4 + 8 \equiv 20 \equiv 2 \ (\text{mod } 9).$$

The check digit on the presumed money order is 3, and therefore it cannot be authentic. ∎

The check digit scheme for U.S. Postal Service money orders will not detect the error of replacing a 0 by a 9 or vice versa in the main part of the serial number because doing so would still result in the sum of the digits of the main part being congruent to the check digit. This kind of error—where one digit is replaced by a different digit in either the main part or check digit—is called a **single-digit error**. The money order check digit scheme will detect any single-digit error except replacing a 0 by a 9 or vice versa. Unfortunately, other kinds of errors would go undetected. For instance, suppose the serial number for the money order with serial number 19187790107 is typed in as 19178790107, with the 87 typed in incorrectly as 78. This kind of error—where two digits are interchanged—is called a **transposition error**. Because every number is congruent to the sum of its digits modulo 9, the only transposition errors that would be detected are those involving the check digit itself. Furthermore, any rearrangement of the digits in the main part of the serial number would still result in the same check digit, and therefore would not be detected.

The U.S. Postal Service money order check digit method detects about 98.0% of the possible single-digit errors but only about 10.0% of possible errors consisting of the transposition of adjacent digits. One study of the frequency of certain types of errors committed by humans when working with identification numbers found that 79.1% of the errors that occur are single-digit errors, whereas only 10.2% of the errors are transpositions of adjacent digits. It would be fair to assume that when identification numbers are read by a scanner rather than a human, single-digit errors account for an even higher percentage of the errors because a machine is unlikely to transpose or rearrange digits. It would therefore be nice to have a check digit scheme that would detect all single-digit errors. The next three methods we will look at all do this.

The check digit method employed by the U.S. banking system in assigning routing numbers to banks detects all single-digit errors and about 88.9% of errors consisting of the transposition of adjacent digits. Every U.S. bank has a nine-digit routing number, which we write as

$$a_8 a_7 a_6 a_5 a_4 a_3 a_2 a_1 a_0,$$

where $a_8 a_7 a_6 a_5 a_4 a_3 a_2 a_1$ is the main part and a_0 is the check digit. In this case the check digit a_0 is given by the following formula:

Bank Routing Number Check Digit Formula

$$a_0 = (7a_8 + 3a_7 + 9a_6 + 7a_5 + 3a_4 + 9a_3 + 7a_2 + 3a_1) \bmod 10$$

The bank routing number on personal checks is the first number on the bottom left. For example, the routing number for the Discover Bank check in Figure 5.2 is 031100649. In this case, the main part of the routing number is 03110064 and the number 9 is the check digit. To see that it is indeed given by our check digit formula, let's compute the check digit ourselves. In doing calculations modulo 10, it is helpful to use the fact that because the remainder of a positive integer when divided by 10 is just its last digit, *a positive integer is congruent to its last digit mod 10*. For the number 031100649, we have

$$a_0 \equiv 7 \cdot 0 + 3 \cdot 3 + 9 \cdot 1 + 7 \cdot 1 + 3 \cdot 0 + 9 \cdot 0 + 7 \cdot 6 + 3 \cdot 4$$
$$\equiv 9 + 9 + 7 + 42 + 12 \equiv 9 + 9 + 7 + 2 + 2 \equiv 29 \equiv 9 \pmod{10},$$

so the check digit should be 9.

126

DATE_____

62-64/311

PAY TO THE
ORDER OF_____ $\$$

_____ DOLLARS

DISCOVER®
BANK
P.O. BOX 30417
SALT LAKE CITY, UT 84130

DISCOVER MONEY MARKET ACCOUNT
800-347-7000

MEMO_____ MP

⑆031100649⑆

FIGURE 5.2 Bank check with routing number 031100649.

The following example gives an instance of a transposition error that is not caught by the U.S banking system check digit.

EXAMPLE 2 *Bank Check*

The routing number of Wells Fargo in Arizona is 122105278. Suppose it is accidentally typed in as 122105728. Will the transposition error of replacing the 27 in the correct number by 72 be detected by the check digit?

SOLUTION:

The check digit for the incorrect number would be

$$a_0 \equiv 7 \cdot 1 + 3 \cdot 2 + 9 \cdot 2 + 7 \cdot 1 + 3 \cdot 0 + 9 \cdot 5 + 7 \cdot 7 + 3 \cdot 2$$
$$\equiv 7 + 6 + 18 + 7 + 45 + 49 + 6$$
$$\equiv 7 + 6 + 8 + 7 + 5 + 9 + 6 \equiv 48 \equiv 8 \pmod{10},$$

so the check digit of 8 is correct for the mistyped number. Therefore, the error will not be detected. ∎

Another check digit method we look at here is the method used in the UPC bar codes found on retail items. The UPC (Universal Product Code) bar code first came into use on grocery store items in 1973 and can now be found on almost all retail items. In Figure 5.3 we have an example of this very familiar bar code.

FIGURE 5.3 UPC bar code for Kellogg's® Crispix®.

Most UPC numbers consist of 12 digits, which are usually written as numbers as well as translated into bars that can be read by a laser scanner. The last digit, a_0, of a 12-digit UPC number, $a_{11}a_{10}a_9a_8a_7a_6a_5a_4a_3a_2a_1a_0$, is a check digit given by the following formula:

UPC Number Check Digit Formula

$$a_0 = -(3a_{11} + a_{10} + 3a_9 + a_8 + 3a_7 + a_6 + 3a_5 + a_4 + 3a_3 + a_2 + 3a_1) \bmod 10.$$

The idea behind this formula is that the check digit a_0 is chosen so that

$$(3a_{11} + a_{10} + 3a_9 + a_8 + 3a_7 + a_6 + 3a_5 + a_4 + 3a_3 + a_2 + 3a_1 + a_0) \bmod 10 = 0.$$

This check digit method detects all single-digit errors and about 88.9% of errors consisting of the transposition of adjacent digits.

Consider the UPC number for a 31.4-ounce box of Crispix® cereal shown in Figure 5.3. Using the UPC number check digit formula, we see that the check digit 9 was given by the calculation

$$a_0 \equiv -(3 \cdot 0 + 3 + 3 \cdot 8 + 0 + 3 \cdot 0 + 0 + 3 \cdot 9 + 9 + 3 \cdot 2 + 2 + 3 \cdot 0)$$
$$\equiv -(3 + 24 + 27 + 9 + 6 + 2) \equiv -(3 + 4 + 7 + 9 + 6 + 2)$$
$$\equiv -31 \equiv -31 + 4 \cdot 10 \equiv 9 \ (\bmod \ 10).$$

EXAMPLE 3 *UPC Number*

The UPC number for a $10\frac{3}{4}$-ounce can of Campbell's® Cream of Mushroom Soup is 051000012616 as shown in Figure 5.4. If the scanner accidentally reads the digit 2 as a 3, will the check digit detect the error?

FIGURE 5.4 UPC bar code for Campbell's® Cream of Mushroom Soup.

SOLUTION:

If the main part of the UPC number is read as 05100001361, then the scanner will expect to see the check digit 3 because the check digit formula gives

$$a_0 \equiv -(3 \cdot 0 + 5 + 3 \cdot 1 + 0 + 3 \cdot 0 + 0 + 3 \cdot 0 + 1 + 3 \cdot 3 + 6 + 3 \cdot 1)$$
$$\equiv -(5 + 3 + 1 + 9 + 6 + 3) \equiv -27 \equiv -27 + 3 \cdot 10 \equiv 3 \ (\bmod \ 10).$$

However, the correct check digit is 6, so this error would be detected. ■

Of the three check digit methods we have looked at, none of them detect all single-digit errors and transposition errors, but there are methods available that do so. One such method was the one used for 10-digit International Standard Book Numbers (ISBN numbers). The check digit for 10-digit ISBN numbers comes from a formula similar to the ones for the UPC number and the U.S. banking system, except that the arithmetic is done modulo 11. Unfortunately, this allows the two-digit number 10 as a possible check digit. However, this problem is easily remedied by using an X to represent the check digit 10. The ISBN number is a 10-digit number $a_9a_8a_7a_6a_5a_4a_3a_2a_1a_0$, where the last digit a_0 is a check digit computed using the following formula:

10-Digit ISBN Check Digit Formula

$$a_0 = (a_9 + 2a_8 + 3a_7 + 4a_6 + 5a_5 + 6a_4 + 7a_3 + 8a_2 + 9a_1) \bmod 11$$

For example, the ISBN number of the paperback version of John Grisham's novel *The Rainmaker* is 0-440-22165-X. The check digit is X because for this ISBN number the check digit formula gives

$$a_0 \equiv 0 + 2 \cdot 4 + 3 \cdot 4 + 4 \cdot 0 + 5 \cdot 2 + 6 \cdot 2 + 7 \cdot 1 + 8 \cdot 6 + 9 \cdot 5$$
$$\equiv 8 + 12 + 10 + 12 + 7 + 48 + 45 \equiv 142 \equiv 10 \pmod{11}.$$ ■

EXAMPLE 4 *10-Digit ISBN Number*

The main part of the 10-digit ISBN number for *Green Eggs and Ham* by Dr. Seuss is 0-394-80016. What is the check digit?

SOLUTION:

The check digit is given by

$$a_0 \equiv 0 + 2 \cdot 3 + 3 \cdot 9 + 4 \cdot 4 + 5 \cdot 8 + 6 \cdot 0 + 7 \cdot 0 + 8 \cdot 1 + 9 \cdot 6$$
$$\equiv 6 + 27 + 16 + 40 + 8 + 54 \equiv 151 \equiv 8 \pmod{11}.$$

Therefore, the check digit is 8. ■

ISBN numbers assigned after January 1, 2007 have 13-digits. The check digit scheme for 13-digit ISBN numbers is a mod 10 scheme similar to the one used for UPC

branching OUT

THE UPC BAR CODE

The 12-digit UPC number of a product is made up of four parts as shown in the UPC number for a box of Q-tips® at right.

The first digit represents the type of product. For instance, a 3 is used for drugs and some other health products. The next five digits are the manufacturer number. In the case of the Q-tips®, the manufacturer is Chesebrough-Pond's USA Co. The five digits following the manufacturer's number are the product number. It is assigned by the manufacturer and identifies the product precisely, including such details as size or weight. The last digit, as we know, is the check digit. To translate the digits of a UPC number into a bar code, the digits are first encoded as seven-digit numbers consisting of only 0's and 1's. This kind of code is called a binary code. The code used for the first digit and the manufacturer's number is the same, whereas the code used for the product number and the check digit can be obtained from the code for the manufacturer's numbers by replacing each 0 by a 1 and each 1 by 0 as shown in the table.

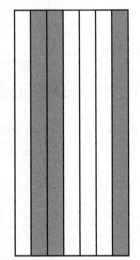

Left-hand Guard Bar Pattern (101)　Center Bar Pattern (01010)　Right-hand Guard Bar Pattern (101)

3　05215　07000　9

Type of Product　Manufacturer's Number　Product Number　Check Digit

Q-tips® UPC bar code.

Image courtesy of the author.

Using the coded representation, each digit is represented in the bar code by seven spaces, called modules, of equal width corresponding to the seven digits of the binary code. The spaces corresponding to 1's are dark and those corresponding to 0's are light as shown in the figure below.

UPC Coding

Digit	Manufacturer's Number and First Digit	Product Number and Check Digit
0	0001101	1110010
1	0011001	1100110
2	0010011	1101100
3	0111101	1000010
4	0100011	1011100
5	0110001	1001110
6	0101111	1010000
7	0111011	1000100
8	0110111	1001000
9	0001011	1110100

UPC bar coding of 0110001.

An entire UPC bar code starts with a left-hand guard bar, three modules wide, corresponding to the binary digits 101. Therefore, it is a light module surrounded by two dark modules. Similarly, it ends with a right-hand guard bar of the same type. These guard bar patterns define the thickness of a module for the scanner. The bars representing the manufacturer's number and the product number are separated by a center bar pattern, five modules wide, corresponding to the binary digits 01010. In addition, each bar code must have margins at least 11 modules wide on the left and the right. Because the binary codes used for the manufacturer numbers have an odd number of 1's, whereas the codes used for the product numbers have an even number of 1's, the scanner will be able to detect whether the code is being scanned from left to right or vice versa, and therefore scanning can be done in either direction.

numbers. This new ISBN scheme does not require use of the letter X, but it no longer has the feature of catching all transposition errors the way that the 10-digit ISBN check digit scheme did.

You might wonder whether it is possible to find a check digit method that detects all single-digit and transposition errors and only generates 0 through 9 as check digits. In 1969 a method satisfying these criteria was discovered by J. Verhoeff. This method uses ideas from an area of mathematics known as group theory.

There are many other methods available for computing check digits. In the exercises, you will look at two other methods commonly used today.

EXERCISES / FOR SECTION 5.4

1. A U.S. Postal Service money order has serial number with main part 6486745669. What is the entire 11-digit serial number?
2. Suppose the serial number for a U.S. Postal Service money order is 57923518026. If the number is accidentally recorded as 57923158026, will the check digit detect the error?
3. Suppose the serial number on a U.S. Postal Service money order is 89012791172. Could it be authentic?
4. The main part of the nine-digit routing number of Bank of America in South Carolina is 05390448. What is the entire bank identification number?
5. The routing number of the First National Bank of Northfield, Minnesota (once robbed by the Jesse James gang), is 091901477. Suppose it is accidentally typed in as 091907417. Will the error be detected by the check digit?
6. Suppose the routing number on a check is 100646023. Could it be an authentic check from a U.S. bank?
7. Suppose the routing number on a valid U.S. bank check is 025500#16, where # is a digit that is illegible. What must the digit # be?
8. The UPC number for the 30-ounce box of Nestle® Quik is 028000242800. If the scanner accidentally reads the digit 4 as a 9 instead, what check digit will the scanner expect to see?

9. The UPC number for a $10\frac{1}{2}$-ounce bag of Kraft® Miniature Marshmallows is 021000660728. If the scanner accidentally reads the digit 7 as a 9, what check digit will the scanner expect to see?

10. The UPC number for a pack of two 9-volt Energizer® batteries is 039800036780. If the number is accidentally entered as 039800036870, will the check digit detect the error?

11. The UPC number for a box of Contac® cold medicine is 345800238209. If the number is accidentally entered as 345800283209, will the check digit detect the error?

12. Suppose you know that the UPC number of a certain product is 3008108#0249, where # is a digit that is illegible. What must the digit # be?

13. The main part of the 10-digit ISBN number of *The Wizard of Oz* by L. Frank Baum is 0-688-06944. Find the check digit.

14. The main part of the 10-digit ISBN number for *Webster's II New Riverside Dictionary* is 0-395-33957. Find the check digit.

15. Suppose you know that the 10-digit ISBN number of a certain book is 0-590-5#880-9, where # is a digit that is illegible. What must the digit # be?

For Exercises 16–19, use the fact that the serial numbers on American Express travelers cheques are 10-digit numbers $a_9a_8a_7a_6a_5a_4a_3a_2a_1a_0$, where $a_9a_8a_7a_6a_5a_4a_3a_2a_1$ is the main part and a_0 is a check digit given by the formula

$$a_0 = -(a_9 + a_8 + a_7 + a_6 + a_5 + a_4 + a_3 + a_2 + a_1) \bmod 9.$$

16. If the main part of the serial number of an American Express travelers cheque is 417948919, find the entire 10-digit serial number.

17. Suppose the serial number of a document that appears to be an American Express travelers cheque is 2893459113. Could it be authentic?

18. The serial number of an American Express travelers cheque is 2570113098.
    (a) If the number is accidentally recorded as 2670113098, will the check digit detect the error?
    (b) Is there any kind of single-digit error that would not be detected by the check digit?

19. The serial number on the American Express travelers cheque is 4370633811.
    (a) If the number is accidentally recorded as 4730633811, will the check digit detect the error?
    (b) Is there any kind of transposition error that will be detected by the check digit?

For Exercises 20–24, consider the following: Some identification codes use both digits and letters. One such type of identification number, known as Code 39, is used by the U.S. Department of Defense, automotive companies, and the health industry. Code 39 identification codes are composed of digits and the 26 uppercase letters A through Z. The identification codes have 15 characters, $a_{14}a_{13}a_{12}a_{11}a_{10}a_9a_8a_7a_6a_5a_4a_3a_2a_1a_0$ where the last character a_0 is a check "character" computed as follows: Any of the main part characters a_k that are letters A through Z are converted to the numbers 10 through 35, respectively. In other words, any A's are converted to 10's, B's are converted to 11's, and so on. The check character is then given by the formula

$$a_0 = -(15a_{14} + 14a_{13} + 13a_{12} + 12a_{11} + 11a_{10} + 10a_9 + 9a_8$$
$$+ 8a_7 + 7a_6 + 6a_5 + 5a_4 + 4a_3 + 3a_2 + 2a_1) \bmod 36,$$

and if a_0 is between 10 and 36, it is converted to the equivalent letter.

20. If the main part of a Code 39 identification code is 34T1CR9244YD3W, find the check character.

21. If the main part of a Code 39 identification code is V97MKU357H8PQB, find the check character.

22. If the Code 39 identification code T313HK21R7732Y5 is typed in as T313HK21R7723Y5, will the check digit detect the error?

23. If the Code 39 identification code KPT26853NT993NR is typed in as JPT26853NT993NR, will the check digit detect the error?

24. Explain why the bank identification number check digit formula may be written alternatively as

$$a_0 = -(3a_8 + 7a_7 + a_6 + 3a_5 + 7a_4 + a_3 + 3a_2 + 7a_1) \bmod 10.$$

5.5 TOURNAMENT SCHEDULING

Modular arithmetic can be used to set up a schedule for a round-robin tournament. In a **round-robin tournament**, each team must play every other team exactly once in a series of rounds. For instance, in the preliminary round of the Olympic Men's Hockey tournament, groups of four teams are scheduled to play round-robin tournaments. In the preliminary round of the 2010 Olympics, Canada, Norway, Switzerland, and the United States played in one group as shown in Table 5.1.

Image © photographer2222, 2012. Shutterstock, Inc.

TABLE 5.1 Round-Robin Schedule in Olympic Men's Hockey

	First Round, February 16	Second Round, February 18	Third Round, February 20 & 21
Canada	Norway	Switzerland	United States
Norway	Canada	United States	Switzerland
Switzerland	United States	Canada	Norway
United States	Switzerland	Norway	Canada

If there are an even number of teams in a round-robin tournament, then every team should be playing in every round. So, for instance, with six teams, it would take five rounds to complete the tournament. If there are an odd number of teams, then in every round one team would have to be scheduled to sit out with a **bye**, and the remaining teams would all play. Therefore, for example, with seven teams in the tournament, there would have to be seven rounds.

Setting up a round-robin tournament for a fairly small number of teams can be done by trial and error because there are relatively few choices to make. However, for a large number of teams it is quite difficult to avoid all of the possible conflicts. (If you are not convinced of this, try to schedule a round-robin tournament even with just 10 teams.) In order to discover a general method that works for any number of teams, we will try to create a schedule with a regular pattern. An irregular construction is unlikely to always work, and if it does, we would have trouble describing it precisely. It turns out to be easier to discover a pattern when we start with an odd number of teams, so we will start with five teams. We give each of the teams a number 1 through 5 as a name. To complete the tournament schedule, we need to schedule five rounds, with one team not playing in each round. The table entry will be the team's opponent for that round. To begin, notice that we can rearrange the rounds of any schedule so that Team 1 plays Teams 2, 3, 4, 5 in order and ends with its bye. We have started such a schedule in Table 5.2, also filling in the schedule for those teams scheduled to play Team 1. This idea illustrates a useful problem-solving technique of reducing the scope of the problem in a way that allows all cases to be constructed from the restricted cases. In our situation, we could find more general schedules by reordering the rounds.

TABLE 5.2 Start of a Round-Robin Schedule for Five Teams

	Round				
	First	Second	Third	Fourth	Fifth
Team 1	2	3	4	5	Bye
Team 2	1				
Team 3		1			
Team 4			1		
Team 5				1	

We must now decide how to proceed; there are many choices. One of the nicest patterns would be if we could, roughly speaking, increase the number of the opponent by one each round. Let's try this for Team 2 and see how it goes, which we have done in Table 5.3.

TABLE 5.3 Part of a Round-Robin Schedule for Five Teams

	Round				
	First	**Second**	**Third**	**Fourth**	**Fifth**
Team 1	2	3	4	5	Bye
Team 2	1	2?	3	4	5
Team 3		1	2		
Team 4			1	2	
Team 5				1	2

Notice that a nice pattern appears to be forming, but we have the problem that Team 2 is scheduled to play itself in round 2. However, the obvious solution is to give Team 2 a bye here. With the optimism spurred by our initial success, we fill in the rest of the schedule by extending this pattern. Whenever a team is scheduled to play itself, it gets a bye. This works out to give the complete round-robin tournament schedule shown in Table 5.4.

TABLE 5.4 Round-Robin Schedule for Five Teams

	Round				
	First	**Second**	**Third**	**Fourth**	**Fifth**
Team 1	2	3	4	5	Bye
Team 2	1	Bye	3	4	5
Team 3	5	1	2	Bye	4
Team 4	Bye	5	1	2	3
Team 5	3	4	Bye	1	2

Careful examination of Table 5.4 may remind you somewhat of an addition table. The way it "wraps around" (see the rows for Teams 3 and 5 in particular) suggests modular arithmetic, possibly arithmetic mod 5 because only five possible values exist. We let $T_{m,r}$ be the team assigned to play Team m in round r. Both m and r can take on any value from 1 through 5. Then in round r, the entry in Table 5.4 is given by $T_{m,r} \equiv r + 2 - m \pmod 5$. We give a bye to the team scheduled to play itself. It is easy to add a sixth team; simply schedule it to play the team with a bye, as in Table 5.5.

TABLE 5.5 Round-Robin Schedule for Six Teams

	Round				
	First	**Second**	**Third**	**Fourth**	**Fifth**
Team 1	2	3	4	5	6
Team 2	1	6	3	4	5
Team 3	5	1	2	6	4
Team 4	6	5	1	2	3
Team 5	3	4	6	1	2
Team 6	4	2	5	3	1

Does such a procedure work in general? It is not obvious that it does because there are several potential pitfalls. In describing a general scheduling procedure, we will simplify slightly by replacing $r + 2$ with r in our formula for $T_{m,r}$. The only effect of this change is to reorder the rounds. Suppose there are N teams in a round-robin tournament, with N odd. We need to set up N rounds. In round r, we begin by letting $T_{m,r}$ be the number between 1 and N such that

$$T_{m,r} \equiv r - m \ (\text{mod } N).$$

For example, if $N = 7$, then in the third round Team 1 will be assigned to play Team 2, because

$$T_{2,3} \equiv 3 - 2 \equiv 1 \ (\text{mod } 7).$$

Similarly, in the third round we have

$$T_{3,3} \equiv 3 - 3 \equiv 0 \equiv 7 \ (\text{mod } 7).$$

so Team 3 will play Team 7. Notice that we could not let $T_{3,3} \equiv 0$ because all of the $T_{m,r}$ must lie between 1 and 7. That is why we adjusted the number 0 by adding 7 in the last step in finding $T_{3,3}$.

The formula we have for $T_{m,r}$ does give some sort of team assignments, but at this point we do not know if the assignments it gives correspond to a round-robin tournament.

One thing that will need to be true for the team assignments given by $T_{m,r}$ is that if Team B is assigned to play Team A in a particular round, then Team A is also assigned to play Team B in that round. To check that this is the case, suppose that Team B is assigned to play Team A. Then it must be true that

$$T_{A,r} \equiv r - A \equiv B \ (\text{mod } N).$$

If we add A and subtract B from both sides of the congruence equation $r - A \equiv B$ (mod N), we get

$$r - A + A - B \equiv B + A - B \ (\text{mod } N),$$

and therefore

$$r - B \equiv A \ (\text{mod } N).$$

Thus, $T_{B,r} \equiv r - B \equiv A \ (\text{mod } N)$, so Team A is assigned to play Team B. Similar computations using modular arithmetic show that in a given round r, different values m in $T_{m,r} \equiv r - m \ (\text{mod } N)$ will give different values of $T_{m,r}$, so two different teams are not assigned to the same opponent in the same round. And in a like way you can show that no team is assigned to play the same team in two different rounds. These points are covered in exercises.

Although it looks like this scheme might work, we have another potential problem. With an odd number of teams, there must be one and only one team assigned to play itself.

Because we have paired off N teams in each round and N is odd, it must be that in each round we have assigned at least one team to play itself. Could more than one team be assigned a bye in a given round? We leave it as an exercise to show that only one team has been assigned to play itself. With an even number of teams, we schedule the first $N-1$ teams in this way and let the team that was assigned to play itself play Team N instead.

We can summarize our method for setting up a round-robin tournament schedule as follows:

Round-Robin Scheduling Method

Suppose the number of teams in the tournament is N, with N odd. Then for $m = 1, 2, \ldots, N$, in round r Team $T_{m,r}$ is assigned to play Team m, where $T_{m,r}$ is the number between 1 and N such that

$$T_{m,r} \equiv r - m \ (\text{mod } N),$$

unless $T_{m,r} = m$, in which case Team m is assigned a bye. When the number of teams is N, with N even, we schedule teams 1 through $N-1$ as for a tournament with $N-1$ teams, except that a team that would be assigned a bye is instead assigned to play Team N.

EXAMPLE 1 *Round-Robin Tournament for Seven Teams*

Using our round-robin scheduling method, set up a round-robin tournament schedule for seven teams.

SOLUTION:

We have $N = 7$, which is odd. In the first round, Team $T_{1,1}$ is assigned to play Team 1, where

$$T_{1,1} \equiv 1 - 1 \equiv 0 \equiv 7 \ (\text{mod } 7).$$

So Team 7 is assigned to play Team 1, and we know, without computing $T_{7,1}$, that Team 1 will be assigned to play Team 7. Next, we see that Team $T_{2,1}$ is assigned to play Team 2, where

$$T_{2,1} \equiv 1 - 2 \equiv -1 \equiv -1 + 7 \equiv 6 \ (\text{mod } 7),$$

so Teams 2 and 6 play each other. Similarly, we have

$$T_{3,1} \equiv 1 - 3 \equiv -2 \equiv -2 + 7 \equiv 5 \ (\text{mod } 7),$$

and therefore Teams 3 and 5 play each other. Next we see that

$$T_{4,1} \equiv 1 - 4 \equiv -3 \equiv -3 + 7 \equiv 4 \ (\text{mod } 7),$$

so Team 4 has been assigned to play itself and will get a bye in round 1. We also should have known that Team 4 would get a bye even before computing $T_{4,1}$ in the first round because it was the only team left to be assigned. We can now complete the schedule for the first round as in Table 5.6.

TABLE 5.6 Beginning of a Round-Robin Schedule for Seven Teams

	Round						
	First	Second	Third	Fourth	Fifth	Sixth	Seventh
Team 1	7						
Team 2	6						
Team 3	5						
Team 4	Bye						
Team 5	3						
Team 6	2						
Team 7	1						

Moving on to the second round, we see that

$$T_{1,2} \equiv 2 - 1 \equiv 1 \ (\text{mod } 7),$$

so Team 1 will have a bye this round. We next find that

$$T_{2,2} \equiv 2 - 2 \equiv 0 \equiv 7 \ (\text{mod } 7),$$

and therefore Team 2 and Team 7 will play each other. Similarly, because

$$T_{3,2} \equiv 2 - 3 \equiv -1 \equiv 6 \ (\text{mod } 7),$$

we see that Team 3 plays Team 6. This leaves out only Teams 4 and 5, so they must play each other. We have now completed the schedule for the second round. The remaining rounds are computed in a similar manner, and we arrive at the complete schedule in Table 5.7.

TABLE 5.7 Round-Robin Schedule for Seven Teams

	Round						
	First	Second	Third	Fourth	Fifth	Sixth	Seventh
Team 1	7	Bye	2	3	4	5	6
Team 2	6	7	1	Bye	3	4	5
Team 3	5	6	7	1	2	Bye	4
Team 4	Bye	5	6	7	1	2	3
Team 5	3	4	Bye	6	7	1	2
Team 6	2	3	4	5	Bye	7	1
Team 7	1	2	3	4	5	6	Bye

Note that after we have assigned all of the teams for the first six rounds, we can figure out the team assignments for the seventh round simply by assigning to each team the team they have not yet been assigned. However, because the calculations are fairly simple, you may want to go ahead and do them to double-check your work. ■

EXAMPLE 2 *Round-Robin Tournament for Eight Teams*

Using our round-robin scheduling method, set up a round-robin tournament schedule for eight teams.

SOLUTION:

We need to make a schedule as for seven teams and let the team assigned a bye in a given round play the Team 8 instead. For example, when we found

$$T_{4,1} \equiv 1 - 4 \equiv -3 \equiv -3 + 7 \equiv 4 \pmod 7$$

for the first round in Example 1, we would schedule Team 4 to play Team 8 in round 1. For the second round,

$$T_{1,2} \equiv 2 - 1 \equiv 1 \pmod 7,$$

so Team 1 will play Team 8 in that round. Otherwise proceeding just as we did for seven teams in Example 1, we create a round-robin schedule for eight teams, given in Table 5.8.

TABLE 5.8 Round-Robin Schedule for Eight Teams

	Round						
	First	Second	Third	Fourth	Fifth	Sixth	Seventh
Team 1	7	8	2	3	4	5	6
Team 2	6	7	1	8	3	4	5
Team 3	5	6	7	1	2	8	4
Team 4	8	5	6	7	1	2	3
Team 5	3	4	8	6	7	1	2
Team 6	2	3	4	5	8	7	1
Team 7	1	2	3	4	5	6	8
Team 8	4	1	5	2	6	3	7

■

Another issue that may arise in setting up a round-robin tournament is in designating which team will be the home team and which will be the away team for each game. To do this fairly, you would like to assign the teams a roughly equal number of home games. Two possible schemes for doing this are explored in the exercises.

EXERCISES FOR SECTION 5.5

In Exercises 1–8, using our round-robin scheduling method, set up a round-robin tournament for the given number of teams.

1. 9 teams
2. 11 teams
3. 10 teams
4. 12 teams
5. 15 teams
6. 13 teams
7. 16 teams
8. 14 teams

9. For a tournament with 20 teams, use our round-robin scheduling method to find the team that will be assigned to play Team 9 in round 7.

10. For a tournament with 24 teams, use our round-robin scheduling method to find the team that will be assigned to play Team 14 in round 5.

11. For a tournament with 31 teams, use our round-robin scheduling method to find the team that will be assigned to play Team 22 in round 12.

12. For a tournament with 17 teams, use our round-robin scheduling method to find the team that will be assigned to play Team 9 in round 3.

13. For a tournament with 22 teams, use our round-robin scheduling method to determine the round in which Team 7 will play Team 10.

14. For a tournament with 19 teams, use our round-robin scheduling method to determine the round in which Team 8 will have a bye.

15. For a tournament with 17 teams, use our round-robin scheduling method to determine the round in which Team 12 will have a bye.

16. For a tournament with 28 teams, use our round-robin scheduling method to determine the round in which Team 15 will play Team 26.

In many tournaments, each game must have a designated home team and away team. One method for doing this in a round-robin tournament is as follows: In any round, if Team i plays Team j, then

  i. if $i + j$ is even, the team with smallest number i or j is the home team,
  ii. if $i + j$ is odd, the team with largest number i or j is the home team.

In Exercises 17–20, use this method to set up a round-robin tournament with home and away teams designated for the given number of teams.

17. 10 teams
18. 12 teams
19. 15 teams
20. 13 teams

21. Suppose the number of teams, N, in a tournament is odd and the home and away assignments are made using the method of Exercises 17–20. Explain why each team will be assigned $(N - 1)/2$ home games.

22. Suppose the number of teams, N, in the tournament is even and the home and away assignments are made using the method of Exercises 17–20. Explain why each team will be assigned either $N/2$ or $(N - 2)/2$ home games.

Another method for designating home and away teams is as follows: In any round, if Team i plays Team j, where $j > i$, then

  i. if $j - i \leq N/2$, Team i is the home team,
  ii. if $j - i > N/2$, Team j is the home team.

In Exercises 23–26, use this method to set up a round-robin tournament with home and away teams designated for the given number of teams.

23. 10 teams
24. 12 teams
25. 15 teams
26. 13 teams

27. Suppose the number of teams, N, in a tournament is odd and the home and away assignments are made using the method of Exercises 23–26. Explain why each team will be assigned $(N-1)/2$ home games.

28. Suppose the number of teams, N, in the tournament is even and the home and away assignments are made using the method of Exercises 23–26. Explain why each team will be assigned either $N/2$ or $(N-2)/2$ home games.

29. Let N be odd. Suppose that the method for scheduling a round-robin tournament given in this section is altered so that in round r Team $T_{m,r}$ is assigned to play Team m, where $T_{m,r}$ is the number between 1 and N such that

$$T_{m,r} \equiv 2r - m \pmod{N}$$

unless $T_{m,r} = m$, in which case Team m is assigned a bye. Set up a round-robin tournament schedule for seven teams using this method. How does the schedule differ from the one we arrived at in Example 1?

30. Let N be odd. Suppose that the method for scheduling a round-robin tournament given in this section is altered so that in round r Team $T_{m,r}$ is assigned to play Team m, where $T_{m,r}$ is the number between 1 and N such that

$$T_{m,r} \equiv r + 2 - m \pmod{N}$$

unless $T_{m,r} = m$, in which case Team m is assigned a bye. Set up a round-robin tournament schedule for eight teams on the basis of this method. How does the schedule differ from the one we arrived at in Example 2?

31. Let N be odd. When you use the relation $T_{m,r} \equiv r - m \pmod{N}$ to assign teams, explain why the same opponent cannot be assigned to two different teams in the same round.

32. Let N be odd. When you use the relation $T_{m,r} \equiv r - m \pmod{N}$ to assign teams, explain why no team is assigned to play the same team in two different rounds.

33. Let N be odd. When you use the relation $T_{m,r} \equiv r - m \pmod{N}$ to assign teams, explain why exactly one team is assigned to play itself.

34. Show that there is exactly one way to complete the schedule shown in the table to create a round-robin tournament schedule for five teams.

| | *Round* | | | | |
	First	Second	Third	Fourth	Fifth
Team 1	2	3	4	5	Bye
Team 2	1	Bye	3	4	5
Team 3		1	2		
Team 4			1	2	
Team 5				1	2

35. Show that there is exactly one way to complete the schedule shown in the table to create a round-robin tournament schedule for eight teams.

	First	Second	Third	Fourth	Fifth	Sixth	Seventh
Team 1	2	3	4	5	6	7	8
Team 2	1	8	3	4	5	6	7
Team 3		1	2	8	4		
Team 4			1	2	3		
Team 5				1	2		
Team 6					1	2	
Team 7						1	2
Team 8		2		3			1

Round

5.6 INTRODUCTION TO CRYPTOLOGY

From the time that people first began to communicate through writing to the present day, there has been an interest in sending written messages in secret codes. Before the advent of postal systems, messages were often sent by private messenger. However, because the messenger could be captured or could be disloyal, it was in the interest of the sender to have the message in secret code. With the widespread use of electronic communication today and the ability of others to intercept these communications, the need for secret codes has become an even more important issue. Number theory plays a key role in many kinds of secret codes.

Before we look at specific secret codes, we will introduce some terminology. The entire discipline of encoding and decoding secret messages is known as **cryptology**. A method for encoding messages is a **cipher**. We call a message that is to be encoded **plaintext**. After the message has been encoded in secret code, we will refer to it as **ciphertext**. The process of encoding a message is sometimes called **enciphering** or **encryption**, and decoding a message is called **deciphering** or **decryption**.

We will limit our discussion of cryptology to the case of encoding and decoding text consisting only of the 26 letters of the alphabet. All of the methods we will consider can be easily adapted to larger character sets including, for instance, numbers, punctuation, or blank spaces.

The Caesar Cipher

One of the earliest known ciphers was used by Julius Caesar and is now called the **Caesar cipher**. With the Caesar cipher, each letter of the message is replaced by the letter three places beyond it in the alphabet, with the last three letters of the alphabet shifted to the first three letters of the alphabet. Caesar used the Roman alphabet, but we will adapt this cipher to the English alphabet. To illustrate the encoding of a message with the Caesar cipher, suppose we start with the plaintext message

I AM IN THE YARD.

Because L is three letters beyond I in the alphabet, we replace I by L. We replace A by D because D is three letters beyond A. Similarly, we replace the remaining letters, and we get the following ciphertext message:

L DP LQ WKH BDUG.

Notice that the Y in the plaintext message was replaced by a B, because in going three spaces forward, A and B follow Z. One problem with this method of enciphering jumps to mind. Because the letter L in the ciphertext stands for a one-letter word, a person trying to decipher the message would already know that the L must stand for A or I because they are the only one-letter words in the English language. Similarly, if the message were longer, the word THE might appear many times in the plaintext, and therefore the word WKH would appear often in the ciphertext. This would give a person trying to decipher the message a strong hint that the ciphertext word WKH is a fairly common word in the English language. Other hints based on particular words that appear in the ciphertext might provide enough information to make the message easy to decipher. Therefore, it is wise to eliminate this problem by breaking the original plaintext message into blocks of letters of the same length. For example, we could break our plaintext message above into blocks of length five and get

IAMIN THEYA RD.

Notice, of course, that the last word was not of length five because we ran out of letters. Enciphering this plaintext message with the Caesar cipher will now give

LDPLQ WKHBD UG.

We have still not eliminated the problem that a very common sequence of letters such as THE in the English language will also appear as a common sequence of letters in the enciphered message (possibly across two blocks). We will see ways to deal with this problem when we look at more complex methods of enciphering in the next section.

We can describe the Caesar cipher and many other ciphers in terms of modular arithmetic. In each of these ciphers, after breaking the letters into blocks, we translate the plaintext message into the integers between 0 and 25 as in Table 5.9.

TABLE 5.9 Numerical Equivalents of Letters

A	B	C	D	E	F	G	H	I	J	K	L	M	N	O	P	Q	R	S	T	U	V	W	X	Y	Z
0	1	2	3	4	5	6	7	8	9	10	11	12	13	14	15	16	17	18	19	20	21	22	23	24	25

With the Caesar cipher, after the plaintext is converted to numbers, we replace each number in the plaintext, P, by a number in ciphertext, C, given by the following formula:

Enciphering Formula for the Caesar Cipher

$$C = (P + 3) \bmod 26$$

This formula gives a value for C between 0 and 25. After converting each number in plaintext to a number in ciphertext with this formula, we then convert the enciphered message back to letters. We see how this whole process works in the following example.

EXAMPLE 1 *Encoding with the Caesar Cipher*

Encipher the following message by breaking it into five-letter blocks and using the Caesar cipher:

MY OFFER IS FIRM.

First we break it into blocks and get

MYOFF ERISF IRM.

Now, converting the letters to their numerical equivalents we get

12 24 14 5 5   4 17 8 18 5   8 17 12.

We now use the Caesar cipher formula $C = (P + 3) \bmod 26$ to transform each number. For instance, for the first number, 12, we get

$$C \equiv 12 + 3 \equiv 15 \ (\bmod \ 26),$$

so $C = 15$. Similarly, the second number, 24, is transformed to 1 because

$$C \equiv 24 + 3 \equiv 27 \equiv 1 \ (\text{mod } 26).$$

Transforming all of the numbers in the plaintext message in this way, we get the following ciphertext message:

15 1 17 8 8   7 20 11 21 8   11 20 15.

Now converting the ciphertext numbers to their letter equivalents, we get the enciphered message

PBRII HULVI LUP,

and this is the message that is sent. ■

When the enciphered message reaches the recipient, it must be deciphered. After converting the enciphered message into numbers, the recipient must transform the ciphertext numbers into plaintext numbers. To decipher a message enciphered with the Caesar cipher, we use the fact that the formula for encoding was $C = (P + 3) \bmod 26$. Starting with this equation and subtracting 3 from both sides, we get

$$C \equiv P + 3 \ (\text{mod } 26),$$
$$C - 3 \equiv P \ (\text{mod } 26),$$

and rearranging we have the following formula for P:

Deciphering Formula for the Caesar Cipher

$$P = (C - 3) \bmod 26$$

EXAMPLE 2 *Decoding with the Caesar Cipher*

Decipher the following message, which was enciphered with the Caesar cipher:

BRXFD QVWDU WQRZ.

SOLUTION:

We begin by converting the letters to their numerical equivalents, and we get

1 17 23 5 3   16 21 22 3 20   22 16 17 25.

Next we use the deciphering formula for the Caesar cipher to convert each ciphertext number to a plaintext number. Starting with the first number, 1, we use the decoding formula $P = (C - 3) \bmod 26$ to find that

$$P \equiv 1 - 3 \equiv -2 \equiv 24 \ (\bmod\ 26),$$

so the 1 will be converted to 24. The second number, 17, will be converted to 14 because

$$P \equiv 17 - 3 \equiv 14 \ (\bmod\ 26).$$

Similarly, we convert the rest of the ciphertext numbers into plaintext numbers, obtaining

$$24\ 14\ 20\ 2\ 0 \qquad 13\ 18\ 19\ 0\ 17 \qquad 19\ 13\ 14\ 22.$$

Now, converting these numbers into their letter equivalents we have

YOUCA NSTAR TNOW,

and breaking the letters into words, we see that the original message was

YOU CAN START NOW. ■

You may now be thinking that we have gone to a great deal of trouble with our formulas for the Caesar cipher. After all, when encoding and decoding with this cipher, we are just shifting the letters back and forth by 3. This can easily be accomplished without introducing modular arithmetic. However, as we will see next, the ideas we have just used can be generalized to give us a large number of different ciphers.

Affine Ciphers

We could generalize the idea of the Caesar cipher by trying a transformation of the form

$$C = (P + b) \bmod 26,$$

where b is some integer between 0 and 25. However, this is still just a shift of the letters. To mix things up a bit more, let's instead try the formula

$$C = (aP + b) \bmod 26,$$

where a and b are both integers between 0 and 25. In order for this idea to be of use, we would have to be able to decipher a message that had been encoded with this formula. This means we would have to be able to solve for P. Starting with the enciphering formula and trying to solve for P using the same ideas we use in solving algebraic equations in regular arithmetic, we have

$$C \equiv aP + b \ (\text{mod } 26),$$

$$C - b \equiv aP \ (\text{mod } 26),$$

$$aP \equiv C - b \ (\text{mod } 26).$$

To finish solving for P in this equation in regular arithmetic, we would multiply both sides by a^{-1}. But what is a^{-1} in arithmetic modulo 26? Let's look at this next.

For a number a, if there exists a number x with $0 \le x < n$ such that

$$ax \equiv 1 \ (\text{mod } n),$$

then we call x the **inverse** of a mod n and denote it by $\bar{a}$ mod n or more briefly $\bar{a}$. For example, because

$$3 \cdot 9 \equiv 27 \equiv 1 \ (\text{mod } 26),$$

we see that $\bar{3}$ mod $26 = 9$. It turns out that not all numbers will have an inverse mod n. For instance, if 13 had an inverse, x, mod 26, then we would have

$$13x \equiv 1 \ (\text{mod } 26).$$

Multiplying both sides of this equation by 2, we get

$$26x \equiv 2 \ (\text{mod } 26),$$

and because $26 \equiv 0$ mod 26, we would have

$$0 \equiv 2 \ (\text{mod } 26).$$

This last statement is not true, so 13 must not have an inverse mod 26. In general, we have the following fact:

The integer a has an inverse mod n if and only if

$$\gcd(a, n) = 1.$$

There is a systematic method based on the Euclidean algorithm for finding the inverse of an integer mod n that works well even for large values of n. However, because we are now interested only in finding the inverses of integers mod 26, we will simply find them through trial and error. We first note that we need only look for inverses of those integers a between 0 and 25 for which $\gcd(a, 26) = 1$—in other words, only the numbers that are not divisible by 2 or 13. We have already noted that $\bar{3}$ mod $26 = 9$, and it follows that $\bar{9}$ mod $26 = 3$. We also see that 1 is its own

TABLE 5.10 Inverses mod 26

a	$\bar{a}$ mod 26
1	1
3	9
5	21
7	15
9	3
11	19
15	7
17	23
19	11
21	5
23	17
25	25

inverse mod 26. Next, let's try to find $\bar{5}$ mod 26. We try the numbers between 0 and 25 that are not divisible by 2 or 13, except 1, 3, and 9 for which we already know the inverses, until we find that

$$5 \cdot 21 \equiv 105 \equiv 1 \;(\text{mod } 26).$$

So 5 and 21 are inverses of each other mod 26. Now, to find $\bar{7}$ mod 26, we try the numbers between 0 and 25 that are not divisible by 2 or 13, except 1, 3, 5, 9, and 21, until we find that $\bar{7}$ mod 26 = 15. Continuing in this way, we find all of the inverses mod 26 as given in Table 5.10.

Looking back now at the idea of using the formula

$$C = (aP + b) \bmod 26$$

to encode a message, recall that in order to decode the message we needed to solve the equation

$$aP \equiv C - b \;(\text{mod } 26).$$

Multiplying both sides of this equation by $\bar{a}$ mod 26 gives

$$\bar{a}aP \equiv \bar{a}(C - b) \;(\text{mod } 26),$$

$$P \equiv \bar{a}(C - b) \;(\text{mod } 26).$$

This formula for P works whenever a has an inverse mod 26. Therefore, we have a way of encoding that works. Codes of the form $C = (aP + b)$ mod 26, with a and b both integers between 0 and 25, and gcd $(a, 26) = 1$, are called **affine ciphers**. Next, we summarize what we have discovered about affine ciphers.

Affine Ciphers

If a and b are both integers between 0 and 25, and gcd(a, 26) = 1, then an affine cipher is given by the enciphering formula

$$C = (aP + b) \bmod 26.$$

It can be deciphered using the formula

$$P = (\bar{a}\,(C - b)) \bmod 26.$$

EXAMPLE 3 *Encoding with an Affine Cipher*

Encipher the following message by breaking it into five-letter blocks and using the affine cipher $C = (5P + 12)$ mod 26:

THE TIME IS NOW.

SOLUTION:

Breaking it into five-letter blocks, we have

THETI MEISN OW.

Changing the letters to their numerical equivalents gives

19 7 4 19 8   12 4 8 18 13   14 22.

Now we are ready to use the transformation $C = (5P + 12)$ mod 26. Transforming the first number, 19, into ciphertext, we get

$$C \equiv 5 \cdot 19 + 12 \equiv 107 \equiv 3 \text{ (mod 26)},$$

so $C = 3$. The second letter, 7, is transformed to 21 because

$$C \equiv 5 \cdot 7 + 12 \equiv 47 \equiv 21 \text{ (mod 26)}.$$

Similarly, we transform the remaining numbers and arrive at the ciphertext numbers

3 21 6 3 0   20 6 0 24 25   4 18,

and transforming these into their letter equivalents gives the ciphertext message

DVGDA UGAYZ ES.   ■

EXAMPLE 4 *Decoding with an Affine Cipher*

Decipher the following message, which was enciphered using the affine cipher $C = (17P + 4)$ mod 26:

PTUMI RPHEM PKYIS EW.

SOLUTION:

Converting the ciphertext letters to their numerical equivalents, we have

15 19 20 12 8   17 15 7 4 12   15 10 24 8 18   4 22.

We know that the deciphering formula is $P = \left(\overline{17}(C - 4) \right)$ mod 26. Because $\overline{17}$ mod 26 = 23, we can write the deciphering formula in the form

$$P = (23(C - 4)) \text{ mod 26.}$$

The first ciphertext number, 15, is transformed to 19 because

$$P \equiv 23(15 - 4) \equiv 23 \cdot 11 \equiv 253 \equiv 19 \text{ (mod 26)}.$$

The second ciphertext number, 19, is transformed to 7 because

$$P \equiv 23(19 - 4) \equiv 23 \cdot 15 \equiv 345 \equiv 7 \ (\text{mod } 26).$$

Similarly, we transform the remaining ciphertext numbers, and we find that the entire message becomes

19 7 4 2 14   13 19 17 0 2   19 8 18 14 10   0 24.

Converting to the letter equivalents, we get

THECO NTRAC TISOK AY,

so the original message was

THE CONTRACT IS OKAY. ■

branching OUT

5.5

MECHANICAL CIPHER DEVICES: FROM ANCIENT GREECE TO ENIGMA

AKG-Images

(a) A scytale.

Mechanical devices for encrypting messages go back 2500 years to Sparta and the scytale shown in figure (a). The scytale consists of a rod of a specific diameter. To encode a message, a ribbon is wrapped around the scytale and a message is written horizontally along the rod, usually along a single row. Other rows of nonsense letters are also written. Once the message and other letters are written, the ribbon is removed and delivered. To decode the message, the recipient rewraps the message around a rod of the same diameter as the one used for encoding. Such a system is not very secure because of the regular appearance of meaningful letters, for instance every sixth letter. Also, the code can be broken easily by anyone who obtains the correct rod, which is the complete encoding and decoding device. With more advanced and secure mechanical devices, codes cannot be broken immediately by someone who obtains the device because

an additional key, which corresponds to the way the device is set up for a particular message, is needed.

(b) A British cipher disk (circa 1880).

A device that is used for encryption and decryption of simple ciphers such as the Caesar cipher is the cipher disk, an example of which is shown in figure (b). The cipher disk was invented by the fifteenth-century Italian Leon Battista Alberti. To use the disk, the inside wheel is set to a specific rotation. The plaintext letters are then on the outside with the corresponding ciphertext letters inside. The Captain

Midnight secret decoder ring from the mid-twentieth century was the children's version of this cipher disk. These devices make it easy to encode and decode quickly and minimize errors.

Courtesy of The National Security Agency, http://www.nsa.gov.

(c) Wheel cipher.

In the 1790s, Thomas Jefferson invented a wheel cipher like the one shown in figure (c). This cipher consisted of several different rotating disks, each containing the 26 letters of the alphabet in some scrambled order. To send a message the sender simply lines up the message along any one row and writes down any of the other 25 rows. This process is repeated until the full message is encoded. To decode the message, a recipient, who has an identical wheel cipher with disks placed in the same order as the sender, just lines up the encoded message along some row. Then the encoded message will appear on some other row. This scheme provided a cipher that was much more difficult to break than the simple substitution schemes. This system was rediscovered and used by the U.S. Army in World War I and later by the U.S. Navy.

In 1817, Colonel Decius Wadsworth invented a geared cipher disk, which was manually rotated as each letter was enciphered, thus changing the key with each successive letter. In 1867, Charles Wheatstone invented a very similar, though slightly less secure, device. These manual devices were the forerunners of machines with rotors that automatically changed the key with each letter. The patent application for a rotor-based machine filed in the Netherlands by Hugo Alexander Koch in 1919 led to the development of the infamous

German Enigma machine from World War II shown in figure (d). The interactions between three rotational rotors, a reflecting rotor, and a plugboard that permuted the text before it was enciphered, along with the enormous number of possible keys, made the Enigma largely unassailable to attack by hand. However, during the war, with the help of the Colossus, according to some the world's first electronic computer, and a model of the Enigma built by Polish mathematicians, the British developed methods for discovering Enigma's keys from encoded messages.

Between spies, traitors, and losses on the battlefield, it is not unusual for mechanical devices to fall into the hands of the enemy. An exception was the Japanese PURPLE machine from World War II. Although American cryptanalysts were eventually able to break the code and deduce the workings of the machine, an intact PURPLE machine was never captured. Ironically, just before the bombing of Pearl Harbor, when Japanese security precautions were at their highest and when U. S. codebreakers were able to partially decipher some Japanese messages only through painstaking work, the Japanese embassy in Washington allowed their American custodian to dust and clean the room containing several PURPLE machines.

Image © markrhiggins, 2012. Shutterstock, Inc.

(d) The German Enigma machine.

The affine ciphers we have looked at are not the only possible ciphers that arise from rearranging the letters of the alphabet. In fact, because only 12 of the numbers between 0 and 25 have inverses mod 26, there are only $26 \cdot 12 - 1 = 311$ such ciphers (not counting the "cipher" $C = P$ mod 26). However, there are $26! \approx 4 \cdot 10^{26}$ ways to permute the letters of the alphabet. Even with this very large number of ways of rearranging the letters, ciphers that are based on this simple idea are fairly easy to decipher if the message is reasonably long. The reason for this is that the approximate frequency with which each letter occurs in ordinary English text is well known. A list of frequencies arrived at by computing the frequencies in a large sample of English text is given in Table 5.11.

TABLE 5.11 Frequency (in percent) of Occurrence of Letters in English Text

A	B	C	D	E	F	G	H	I	J	K	L	M	N	O	P	Q	R	S	T	U	V	W	X	Y	Z
8	1.5	3	4	13	2	2	6	7	<1	1	4	2	7	7.5	2	<1	6	6	9	3	1	2	<1	2	<1

This information can be used to attempt to decipher a message that was encoded by a rearrangement of the letters of the alphabet. For instance, if the most frequent letter occurring in a ciphertext message of at least moderate length is R, then there is a good possibility that the corresponding plaintext letter is E. Similarly, letters that occur very infrequently in the ciphertext probably correspond to one of the uncommon letters in Table 5.11. In addition, certain letter combinations can be helpful in decoding. For instance, if the letter T in ciphertext is always followed by an N (possibly across two blocks), then it may be that the corresponding plaintext letters are Q and U. Similarly, if the three-letter sequence YCP occurs often in a ciphertext message, then the corresponding plaintext letters may very well be THE or AND or some other common three-letter sequence. Using these kinds of clues along with the letter frequency data makes codes based on rearranging the alphabet too easy to break if the message is long. In the next section, we will look at ciphers that encode blocks of letters. These kinds of codes can be difficult or nearly impossible to break.

EXERCISES FOR SECTION 5.6

1. Encipher the message

   ANY TIME IS OKAY

   by breaking it into five-letter blocks and using the Caesar cipher.

2. Encipher the message

   THIS IS THE WAY TO GO

   by breaking it into five-letter blocks and using the Caesar cipher.

3. Decipher the following message, which was enciphered with the Caesar cipher:

BRXJR WLWUL JKW.

4. Decipher the following message, which was enciphered with the Caesar cipher:

DZDBZ HJR.

In Exercises 5–8, encode the given message by breaking it into five-letter blocks and using the given affine cipher.

5. YOU SHOULD STAY, $C = (5P + 14) \mod 26$
6. IT IS THE RIGHT SIZE, $C = (11P + 6) \mod 26$
7. HERE IS THE SECRET, $C = (19P + 7) \mod 26$
8. NOW WE KNOW, $C = (9P + 10) \mod 26$

In Exercises 9–12, decode the given message, which was encoded using the given affine cipher.

9. RPQCU TQGQU D, $C = (7P + 16) \mod 26$
10. FMZMI GCMOD ZGSTN M, $C = (15P + 4) \mod 26$
11. ATWEE OLWFE VLDX, $C = (23P + 9) \mod 26$
12. VLPXV FULF, $C = (3P + 11) \mod 26$

13. Suppose that the formula $C = (6P + 5) \mod 26$ is used for encoding.
    (a) Show that the letter G is encoded as P.
    (b) Show that the letter T is also encoded as P.
    (c) Why will the formula $C = (6P + 5) \mod 26$ not work as an affine cipher?
14. Suppose you need a cipher for encoding numbers rather than words (for instance encoding Social Security numbers). Explain how you could adapt the affine cipher we used for letters to get a cipher for encoding numbers.
15. Suppose you need a cipher for encoding messages that include numbers, words, and blank spaces. Explain how you could adapt the affine cipher we used for letters to get a cipher for doing this.

5.7 ADVANCED ENCRYPTION METHODS

We have seen that ciphers that only rearrange the letters of the alphabet can be broken by use of letter frequency or other clues. To make letter frequency data unusable, we need a code that does more than rearrange individual letters. We will now look at ciphers that encode blocks of letters or blocks of digits coming from letters.

In the first cipher we will look at in this section, the Hill cipher, blocks of letters of a certain length in plaintext are replaced by blocks of letters of the same length in ciphertext. For instance, for a particular code the three-letter block THE in plaintext may be replaced by YWP in ciphertext, whereas the block TIO may be replaced by NCJ. Notice that although the T was encoded as Y in the first case, it was encoded as N in the second case. In block codes of this type, each block is always encoded the same way, but each individual letter may be encoded various ways. This prohibits a person who is trying to decode the message from using single-letter frequency data as an aid.

With the second kind of cipher we will look at in this section, the RSA public key system, instead of encoding blocks of letters of a particular length, the code works on blocks of digits of the same length. However, the basic idea is the same as with blocks of letters, and again the code is not simply a rearrangement of the alphabet. As we will see later, the RSA public key system also has other features that make it a particularly attractive method.

The Hill Cipher

The **Hill cipher** was presented by Lester S. Hill in a paper published in 1931. It works for blocks of letters of any size, but for simplicity we will first look at how it works in the case of two-letter blocks.

With the Hill cipher that we will discuss, each block of two plaintext letters P_1P_2 is encoded as a block of two ciphertext letters C_1C_2, which are given by two equations of the form

$$C_1 = (aP_1 + bP_2) \bmod 26,$$

$$C_2 = (cP_1 + dP_2) \bmod 26,$$

where a, b, c, and d are integers between 0 and 25. We will see later that in order for this type of code to work, we must also require that $\gcd(ad - bc, 26) = 1$. Before seeing why this is the case and finding the deciphering formula, let's look at an example of encoding using a Hill cipher for two-letter blocks.

EXAMPLE 1 *Encoding with the Hill Cipher*

Use the Hill cipher for two-letter blocks given by

$$C_1 = (12P_1 + 3P_2) \bmod 26,$$

$$C_2 = (5P_1 + 6P_2) \bmod 26$$

to encipher the message

CALL TODAY.

SOLUTION:

First we break the message into the two-letter blocks and add a "dummy" letter at the end so that the last block is two letters long. Using the dummy letter T (any letter will do) we have

CA LL TO DA YT.

Next, the letters are translated into their numerical equivalents and we get

2 0   11 11   19 14   3 0   24 19.

Now, we use the Hill cipher formula to encode the first block 2 0 with $P_1 = 2$ and $P_2 = 0$ and we get

$$C_1 \equiv 12 \cdot 2 + 3 \cdot 0 \equiv 24 \ (\mathrm{mod}\ 26),$$

$$C_2 \equiv 5 \cdot 2 + 6 \cdot 0 \equiv 10 \ (\mathrm{mod}\ 26).$$

Therefore, the block 2 0 will be encoded as the block 24 10. Similarly, for the second block 11 11, we set $P_1 = 11$ and $P_2 = 11$, and we have

$$C_1 \equiv 12 \cdot 11 + 3 \cdot 11 \equiv 165 \equiv 9 \ (\mathrm{mod}\ 26),$$

$$C_2 \equiv 5 \cdot 11 + 6 \cdot 11 \equiv 121 \equiv 17 \ (\mathrm{mod}\ 26).$$

Thus, the block 11 11 is encoded as 9 17. Continuing in this way, we can encode the remaining three blocks in the plaintext message, and we get the ciphertext message in numerical form:

24 10   9 17   10 23   10 15   7 0.

Translating back to letters, we get the ciphertext message

YK JR KX KP HA.   ■

In an affine cipher, the letter A is always encoded as the same letter. Notice that, under the Hill cipher we just considered, the block CA was encoded as YK, whereas the block DA was encoded as KP. Similarly, the block LL was encoded as JR. This is to be expected because we are not just rearranging the alphabet. Any particular letter may be replaced by any other letter. However, any particular two-letter block will always be encoded the same way.

To see how to decode a message that was encoded with a Hill cipher for two-letter blocks, we start with the encoding formulas

$$C_1 = (aP_1 + bP_2) \ \mathrm{mod}\ 26,$$

$$C_2 = (cP_1 + dP_2) \ \mathrm{mod}\ 26,$$

and we solve for P_1 and P_2 using the same ideas we would use to solve for two unknowns in two equations of this form in regular arithmetic. In order to eliminate the variable P_2, we begin by multiplying the first equation by d on both sides and the second equation by b to get

$$dC_1 \equiv adP_1 + bdP_2 \,(\text{mod } 26),$$

$$bC_2 \equiv bcP_1 + bdP_2 \,(\text{mod } 26).$$

Now subtracting the second equation from the first, it follows that

$$dC_1 - bC_2 \equiv (ad - bc)P_1 \,(\text{mod } 26).$$

Recall that we required that $\gcd(ad - bc, 26) = 1$, and we can now see why we needed to make this restriction. To solve for P_1 in the last equation, we need to multiply both sides by $\overline{ad - bc}$ mod 26. Doing this, we get

$$\left(\overline{ad - bc}\right)(dC_1 - bC_2) \equiv \left(\overline{ad - bc}\right)(ad - bc)P_1 \,(\text{mod } 26),$$

and therefore

$$P_1 \equiv \left(\left(\overline{ad - bc}\right)(dC_1 - bC_2)\right) \text{mod } 26,$$

or

$$P_1 \equiv \left(\left(\overline{ad - bc}\right)dC_1 - \left(\overline{ad - bc}\right)bC_2\right) \text{mod } 26,$$

Letting $A = \left(\left(\overline{ad - bc}\right)d\right) \text{mod } 26$ and $B = -\left(\left(\overline{ad - bc}\right)b\right) \text{mod } 26$, we have

$$P_1 = (AC_1 + BC_2) \text{mod } 26.$$

We leave it as an exercise to show that in a similar way we can find that

$$P_2 = (CC_1 + DC_2) \text{mod } 26,$$

where $C = -\left(\left(\overline{ad - bc}\right)c\right) \text{mod } 26$ and $D = \left(\left(\overline{ad - bc}\right)a\right) \text{mod } 26$. The formulas for encoding and decoding using the Hill cipher for two-letter blocks are summarized next.

The Hill Cipher for Two-Letter Blocks

If a, b, c, and d are integers between 0 and 25 and $\gcd(ad - bc, 26) = 1$, then the Hill cipher for two-letter blocks is given by the enciphering formulas

$$C_1 = (aP_1 + bP_2) \text{ mod } 26,$$
$$C_2 = (cP_1 + dP_2) \text{ mod } 26.$$

The Hill Cipher for Two-Letter Blocks (*continued*)

It can be deciphered using the formulas

$$P_1 = (AC_1 + BC_2) \bmod 26,$$
$$P_2 = (CC_1 + DC_2) \bmod 26,$$

where

$$A = \left(\left(\overline{ad - bc}\right)d\right) \bmod 26,$$
$$B = -\left(\left(\overline{ad - bc}\right)b\right) \bmod 26,$$
$$C = -\left(\left(\overline{ad - bc}\right)c\right) \bmod 26,$$
$$D = \left(\left(\overline{ad - bc}\right)a\right) \bmod 26.$$

Notice that the formulas for encoding and decoding with the Hill cipher are of the same form. In the following example, we look at the decoding process for the Hill cipher.

TECHNOLOGY *tip*

A spreadsheet or mathematical calculator is a great aide in encoding and decoding affine ciphers and the Hill cipher. It is especially easy with a MOD function or key, but it is also possible to build up your own MOD function with an integer part function or key.

EXAMPLE 2 *Decoding with the Hill Cipher*

Decipher the following message:

$$MZ \ EL \ BW \ MQ \ YM \ EZ,$$

which was enciphered using the Hill cipher

$$C_1 = (13P_1 + 2P_2) \bmod 26,$$
$$C_2 = (9P_1 + 1P_2) \bmod 26.$$

SOLUTION:

Converting the ciphertext letters into their numerical equivalents, we have

$$12\ 25 \quad 4\ 11 \quad 1\ 22 \quad 12\ 16 \quad 24\ 12 \quad 4\ 25.$$

To find the deciphering formulas, we first compute

$$ad - bc \equiv 13 \cdot 1 - 2 \cdot 9 \equiv -5 \equiv 21 \ (\text{mod } 26),$$

and then we see that

$$\overline{ad - bc} \bmod 26 = \overline{21} \bmod 26 = 5.$$

In our deciphering formula we have

$$A \equiv 5 \cdot 1 \equiv 5 \ (\text{mod } 26),$$
$$B \equiv -5 \cdot 2 \equiv -10 \equiv 16 \ (\text{mod } 26),$$
$$C \equiv -5 \cdot 9 \equiv -45 \equiv 7 \ (\text{mod } 26),$$
$$D \equiv 5 \cdot 13 \equiv 65 \equiv 13 \ (\text{mod } 26),$$

and therefore

$$P_1 = (5C_1 + 16C_2) \bmod 26,$$
$$P_2 = (7C_1 + 13C_2) \bmod 26.$$

Translating the first block, 12 25, we get

$$P_1 \equiv 5 \cdot 12 + 16 \cdot 25 \equiv 460 \equiv 18 \ (\text{mod } 26),$$
$$P_2 \equiv 7 \cdot 12 + 13 \cdot 25 \equiv 409 \equiv 19 \ (\text{mod } 26),$$

and therefore the first block translates into the block 18 19. Similarly, we translate each of the other two-number blocks and the entire message is translated to

$$18\ 19 \quad 14\ 15 \quad 19\ 7 \quad 4\ 6 \quad 0\ 12 \quad 4\ 15.$$

Now changing the numbers to their numerical equivalents, we get

ST OP TH EG AM EP.

The last letter, P, appears to be a dummy letter, so the original message was

STOP THE GAME. ■

The Hill cipher for two-letter blocks is easy to use and harder to break than the affine codes we looked at in the last section. However, it is far from unbreakable. There are only $26^2 = 676$ different two-letter blocks, and the frequencies of these two-letter blocks in the English language are well known. For instance, according to some studies the most common two-letter block in the English language is TH, and the second most common is HE. This kind of information may provide enough clues to enable a third party to decode a private message. The Hill cipher is much more secure if it is used with larger blocks. The idea we saw with two-letter blocks generalizes easily to larger blocks. For instance, with the Hill cipher for three-letter blocks, each block of three plaintext letters $P_1P_2P_3$ is encoded as a block of three ciphertext letters $C_1C_2C_3$, which are given by the three equations

$$C_1 = (aP_1 + bP_2 + cP_3) \bmod 26,$$
$$C_2 = (dP_1 + eP_2 + fP_3) \bmod 26,$$
$$C_3 = (gP_1 + hP_2 + iP_3) \bmod 26,$$

where a through i are integers between 0 and 25 that satisfy a condition that will make decoding possible. The Hill cipher for blocks of other sizes is similar. In general, the larger the block size, the harder the code is to break. For instance, if we use eight-letter blocks, then because there are $26^8 \approx 200$ billion different eight-letter blocks, trying to decode a message based on the frequency of eight-letter blocks is not really practical. Therefore, the Hill cipher seems to be a good encryption method if used with large blocks. However, if a small portion of the original text is also known, the entire cipher is easily cracked. Because of this major drawback, the Hill cipher is rarely used.

The RSA Public Key System

With the encryption methods we have looked at so far, if the enciphering method is known, then the deciphering method is easy to compute, so the enciphering method must remain private. In the age of electronic communications, when a person may receive private communications from various sources, it would be more convenient if the person could publicly state the encryption method all correspondents should use when sending messages to him or her. This way, all of the person's messages would arrive encoded in the same way. However, you might wonder how an encryption method could be made public without giving away the decoding method at the same time. Remarkably, such encoding systems are possible and are called public key systems.

In 1975, Ron Rivest, Adi Shamir, and Len Adelman invented a public key system, known as the **RSA public key system**. With the RSA system, each person who would like to receive encoded messages chooses a pair of very large prime numbers p and q and a positive integer r such that $\gcd(r, (p-1)(q-1)) = 1$. The person then calculates the product $n = pq$ and publicly announces that his or her RSA **public key address** is (r, n). To send a message to this individual, the sender first converts all of the letters into their numerical equivalents, placing a zero before any one-digit number. For instance, the letter A would be translated as 00 and the letter B would become 01. The sender then breaks the message into blocks of some agreed upon number of digits (not letters). Any size blocks will do as long as they are of size at most one less than the number of digits in n. Then each plaintext block P is encoded as a ciphertext number C by using the formula

$$C = P^r \bmod n.$$

Note that C can be any number between 0 and $n-1$, so it does not necessarily have the same number of digits as P. The magic of this system, as we will see shortly, is that although anyone who knows the address (r, n) can send messages in this way, only a person who knows how to factor the number n as $n = pq$ will be able to decode the message. This might not seem difficult, but if p and q are sufficiently large primes, factoring n may take much more than a lifetime to carry out, even with the help of today's computer technology. Surprisingly, it is relatively easy to find very large primes, so it is easy to come up with public key addresses. Before looking at how we go about decoding a message that has been encoded with this method, let's look at an example of encoding.

EXAMPLE 3 *Encoding with the RSA Public Key System*

Encode the message

THERE IS A DELAY

using three-digit blocks and the RSA public key system. Assume the message is being sent to a person with public key address (3, 2497).

SOLUTION:

We note first that the person encoding the message does not need to know how to factor 2497, although in this case it is fairly easy to see that $2497 = 227 \cdot 11$ is the product of the prime numbers 227 and 11. Also, notice that

$$\gcd(3, (227-1)(11-1)) = \gcd(3, 2260) = 1,$$

so the address (3, 2497) is of the correct form.

To encode the message, we first translate all of the letters into two-digit numbers and get

19 07 04 17 04   08 18   00   03 04 11 00 24.

Regrouping the digits into three-digit blocks, we get

190   704   170   408   180   003   041   100   240.

We added a dummy digit 0 to fill out the last block. Using the encryption formula

$$C = P^3 \bmod 2497$$

to encode the first block $P = 190$, we find that

$$C \equiv 190^3 \equiv 6{,}859{,}000 \equiv 2238 \ (\text{mod } 2497),$$

so the block 190 is encoded as 2238. The second block, $P = 704$, is encoded as 0363 because

$$C \equiv 704^3 \equiv 348{,}913{,}664 \equiv 363 \ (\text{mod } 2497).$$

Notice that we added the digit 0 to the front of 363 to make it a four-digit block. We want all of the encoded blocks to be four digits long (equal to the number of digits in the modulus 2497) so that the message is more uniform and can be sent without including spaces if desired. To ensure this, we may sometimes have to attach zeros to the fronts of the encoded blocks. Continuing in this way, we encode the remaining seven blocks and the entire message is encoded as

2238   0363   1401   1409   1505   0027   1502   1200   0608.

This is the message that is sent to the person with public key address $(3, 2497)$. Notice that we do not convert the encoded message into letters with this encryption system. In fact, we really do not have the option of easily converting to letters at this stage because the two-digit blocks are not all between 0 and 25. ∎

To decode a message that has been encoded using the public key address (r, n), with $n = pq$, the recipient uses the simple formula

$$P = C^k \bmod n,$$

where

$$k = \bar{r} \bmod ((p - 1)(q - 1)),$$

the inverse of r in arithmetic mod $((p - 1)(q - 1))$. The proof that this deciphering method works is based on a famous theorem from number theory known as Euler's Theorem. Finding the value of k can be accomplished through a simple algorithm as long as the value $(p - 1)(q - 1)$ is known. However, recall that the only information available to the public is the address (r, n), and finding the value $(p - 1)(q - 1)$ requires knowing the factorization $n = pq$. Because this factorization is only known by the recipient and factoring n is very difficult if p and q are very large, the encoded message can only be deciphered by the recipient. In practice, the recipient may only know the value k, having purchased the public key address and the decoding key k from a company that knows the factorization $n = pq$. In the next example, we see how this decoding procedure works.

EXAMPLE 4 *Decoding with the RSA Public Key System*

The following message was encoded using three-digit blocks and the public key address (1373, 1817):

$$0684 \qquad 1098 \qquad 0973 \qquad 0803 \qquad 0531 \qquad 1313.$$

The recipient of the message knows that $1817 = 79 \cdot 23$ and

$$\overline{1373} \bmod ((79-1)(23-1)) = \overline{1373} \bmod 1716 = 5.$$

Decode the message.

SOLUTION:

We use the decoding formula

$$P = C^5 \bmod 1817.$$

Note that the plaintext blocks were three digits long, so when using the decoding formula we have to write the resulting P in the form of a three-digit block, possibly by adding zeros to the front of P. For the first block of digits, 0684, we need to evaluate $P = 684^5 \bmod 1817$. Because 684^5 is a fairly large number that a calculator may not be able to compute exactly, we break down the computation into smaller powers of 684 and get

$$P \equiv 684^5 \equiv 684^2 \cdot 684^2 \cdot 684 \equiv (467{,}856)(467{,}856)(684)$$
$$\equiv (887)(887)(684) \equiv 538{,}149{,}996 \equiv 21 \pmod{1817},$$

so the first three-digit block in plaintext is 021. For the second block of digits, 1098, we have

$$P \equiv 1098^5 \equiv 1098^2 \cdot 1098^2 \cdot 1098$$
$$\equiv (1{,}205{,}604)(1{,}205{,}604)(1098)$$
$$\equiv (933)(933)(1098) \equiv 955{,}796{,}922 \equiv 412 \pmod{1817},$$

so the second three-digit block in plaintext is 412. Similarly, the remaining blocks are translated and we get the six three-digit plaintext blocks

$$021 \qquad 412 \qquad 040 \qquad 704 \qquad 170 \qquad 400.$$

Breaking them into the two-digit blocks representing the plaintext letters, we have

$$02 \quad 14 \quad 12 \quad 04 \quad 07 \quad 04 \quad 17 \quad 04 \quad 00,$$

and translating these numbers to their letter equivalents we have

C O M E H E R E A.

The letter A appears to have come from dummy digits, so we can see that the original plaintext message was

COME HERE. ■

The formulas for encoding and decoding using the RSA public key system are summarized next.

The RSA Public Key System

If p and q are prime numbers, $n = pq$, and r is a positive integer with $\gcd(r, (p-1)(q-1)) = 1$, then a message to be sent to the RSA public key address (r, n) is encoded as follows: Each plaintext block P of digits is encoded as a ciphertext block C by using the formula

$$C = P^r \bmod n.$$

The recipient of the message can decode each ciphertext block C using the formula

$$P = C^k \bmod n,$$

where

$$k = \overline{r} \bmod ((p-1)(q-1)).$$

The blocks of plaintext digits must all be of the same designated length, which must be less than the number of digits of n. The blocks of ciphertext digits must be of length equal to the number of digits of n.

As we noted earlier, the security of the RSA public key system is based on the fact that, although it is fairly easy to find large primes p and q, it is difficult to factor the number $n = pq$. For example, it does not take much computer time to find primes p and q with 200 digits each. However, in this case, the product $n = pq$ will have about 400 digits, and factoring a number of this size that has no small factors might take as much as a billion years of computer time. As computers become faster and better algorithms are developed, the time required to factor is expected to decrease significantly, so larger values of p and q may become necessary. With the kind of computers we use today it is never expected to become especially easy to factor such numbers, so the RSA public key system would continue to be a secure enciphering method. However, as described in Branching Out 5.6, a new kind of computer, called a quantum computer, is under development. If realized, these computers would be so fast that factoring large numbers would become a fairly easy task, therefore breaking the RSA code.

Another advantage of the RSA public key system is that if a third party obtains a message in both its plaintext and ciphertext versions, he or she would still not know how the

branching OUT

Privacy and National Security: AES and CRACKING RSA

Image © mkabakov, 2012. Shutterstock, Inc.

The RSA public key system is in widespread use today. Although the RSA system can be used by itself, it is more common to couple RSA with another encryption system such as AES (for Advanced Encryption Standard) which was adopted by the U.S. government in 2001 and is now used worldwide. With AES, messages can be encoded and decoded at a much faster rate than with RSA. However, AES uses the same key for both encryption and decryption, so the keys must be transmitted securely between the parties. This is where RSA comes in. Because RSA is a public key cipher, the sender can use it to transmit the AES key to the recipient. The message itself is encoded with AES using that key. After using the RSA system to decode the key, the recipient is able to decode the AES-encrypted message. At this time AES is considered secure. The U.S. government considers AES key lengths of 128, 192, and 256 bits sufficient to protect classified information up to the SECRET level. For TOP SECRET information, key lengths of 192 or 256

bits are required. A regular decimal digit translates to about 3.32 bits, so a 256 bit key length corresponds to a key length of 77 digits.

How secure is RSA? The only known way to break RSA is to factor the modulus. As of early 2012, the largest number factored by a general-purpose factoring algorithm was 232 digits long. In practice, the number of digits in the modulus $n = pq$ used in the RSA encoding formula is between 300 to 600 digits. However, factoring algorithms and computers continue to get faster, so it is generally recommended that the modulus n be a minimum of 600 digits long.

The greatest challenge to the security of RSA may be the possible development of a working quantum computer. The computers in use today understand bits that can have only the values of 0 or 1. Quantum computers understand quantum bits, also know as qubits, that can have the value of 0 or 1 or both values at the same time. This difference would allow a functioning quantum computer to perform millions of calculations at once, making it amazingly faster than today's computers. A quantum computer would be able to factor the modulus n so fast that it would effectively break the RSA code. In February 2012, IBM announced that it had made significant advances in the development of quantum computers. Although the time line is debatable, many believe that we will see working quantum computers within our lifetime.

decoding is done. Therefore, this third party would be unable to decipher other messages encoded the same way. This would not be the case with the affine cipher or Hill cipher. With these ciphers, having access to even a short message in both its plaintext and ciphertext versions would usually give a third party who knew the general method of encoding enough information to find the exact encoding and decoding formulas.

A CLOSER *look*

RAISING A NUMBER TO A HIGH POWER IN MODULAR ARITHMETIC

When encoding and decoding in the RSA cryptosystem, we must raise a number to an exponent in modular arithmetic. The exponents in our examples were small, but in actual practice these exponents may be quite large. To simplify the calculations, successively squaring is helpful. To illustrate this idea, let's calculate 35^{27} mod 57. First we successively square 35 modulo 57.

$$35^2 \equiv 1225 \equiv 28 \ (\text{mod } 57),$$

$$35^4 \equiv 35^2 \cdot 35^2 \equiv 28 \cdot 28 \equiv 784 \equiv 43 \ (\text{mod } 57),$$

$$35^8 \equiv 35^4 \cdot 35^4 \equiv 43 \cdot 43 \equiv 1849 \equiv 25 \ (\text{mod } 57),$$

$$35^{16} \equiv 35^8 \cdot 35^8 \equiv 25 \cdot 25 \equiv 625 \equiv 55 \ (\text{mod } 57).$$

Now using the fact that $27 = 16 + 8 + 2 + 1$, we have

$$35^{27} \equiv 35^{16+8+2+1} \equiv 35^{16} \cdot 35^8 \cdot 35^2 \cdot 35$$

$$\equiv 55 \cdot 25 \cdot 28 \cdot 35 \equiv 1{,}347{,}500 \equiv 20 \ (\text{mod } 57).$$

Because any positive integer can be written as a sum of powers of two (this is the base 2 expansion), we can apply this technique in general. Performing the calculation in this way keeps down the total number of multiplications to be performed. For instance, if we had computed 35^{27} by successively multiplying by 35, we would have had to perform 26 multiplications. Using the successive squaring method, we performed only seven multiplications. For larger exponents the savings in multiplication are even more notable. For instance, an exponent of one trillion requires only 51 multiplications using the successive squaring method.

E X E R C I S E S /// FOR SECTION 5.7

In Exercises 1–4, encode the given message using the given Hill cipher for two-letter blocks. Use the letter X as a dummy letter in the last block if necessary.

1. A STORM IS COMING; $C_1 = (4P_1 + 3P_2) \bmod 26,$
$$C_2 = (15P_1 + 5P_2) \bmod 26$$
2. GOOD MORNING; $C_1 = (11P_1 + 4P_2) \bmod 26,$
$$C_2 = (3P_1 + 9P_2) \bmod 26$$
3. GO FOR IT; $C_1 = (5P_1 + 1P_2) \bmod 26,$
$$C_2 = (6P_1 + 13P_2) \bmod 26$$
4. PLEASE CALL; $C_1 = (21P_1 + 8P_2) \bmod 26,$
$$C_2 = (1P_1 + 3P_2) \bmod 26$$

In Exercises 5–8, decode the given message, which was encoded using the given Hill cipher for two-letter blocks with some dummy letter in the last block when needed.

5. DF ZH TZ PJ UI; $C_1 = (5P_1 + 2P_2) \bmod 26,$
$C_2 = (3P_1 + 7P_2) \bmod 26$

6. VW ED BC QJ; $C_1 = (1P_1 + 17P_2) \bmod 26,$
$C_2 = (4P_1 + 9P_2) \bmod 26$

7. UB NC SP SB VJ; $C_1 = (6P_1 + 7P_2) \bmod 26,$
$C_2 = (19P_1 + 3P_2) \bmod 26$

8. AR JB DS GD BI AJ; $C_1 = (16P_1 + 5P_2) \bmod 26,$
$C_2 = (5P_1 + 8P_2) \bmod 26$

In Exercises 9–12, encode the given message using three-digit blocks and the RSA public key system where the message is being sent to the person with the given public key address. Use the digit 0 for the dummy digits in the last block if necessary.

9. TELL ME WHY; (3, 3071)
10. SEND MONEY; (3, 4189)
11. I FOUND IT; (5, 7081)
12. GET BACK; (7, 5459)

In Exercises 13–16, decode the given message, which was encoded using the RSA public key system with three-digit blocks and sent to the given address. The digit 0 was used for the dummy digits in the last block if needed.

13. 1472   0012   1099   0944   0038   1098;
Address: (1083, 1711),
where $1711 = 29 \cdot 59$ and $\overline{1083} \bmod ((29-1)(59-1)) = \overline{1083} \bmod 1624 = 3$

14. 0746   0703   1009   0035   0657   0629   1009;
Address: (1027, 1633),
where $1633 = 23 \cdot 71$ and $\overline{1027} \bmod ((23-1)(71-1)) = \overline{1027} \bmod 1540 = 3$

15. 3542   1701   1494   2691;
Address: (2423, 4387),
where $4387 = 41 \cdot 107$ and $\overline{2423} \bmod ((41-1)(107-1)) = \overline{2423} \bmod 4240 = 7$

16. 0998   5408   2751   1719;
Address: (1613, 8249),
where $8249 = 73 \cdot 113$ and $\overline{1613} \bmod ((73-1)(113-1)) = \overline{1613} \bmod 8064 = 5$

17. Suppose that the formulas
$$C_1 = (5P_1 + 3P_2) \bmod 26,$$
$$C_2 = (2P_1 + 9P_2) \bmod 26$$
are used for encoding.
(a) Show that the block of letters SH is encoded as HV.
(b) Show that the block of letters CZ is also encoded as HV.
(c) Why will the formulas not work as a cipher?

18. A Closer Look 5.3 examines an efficient way to calculate large powers in modular arithmetic. A commonly used exponent for the RSA public key system is the prime $65{,}537 = 2^{16} + 1$. Explain why an exponent of this form would yield an easier computation than would another exponent of comparable size.

19. How could the RSA public key system be adapted for sending messages that everyone could read but that only one person could send?

20. How could the RSA public key system be adapted so a sender could add a "signature" to a message that allows the receiver to verify that the message came from the actual sender and is not a message from someone else claiming to be the sender?

WRITING EXERCISES

1. Write a letter to the head of your college supporting the use of check digits in student numbers.
2. Discuss some of the practical issues faced by those who create league schedules.
3. Explain important applications of public key ciphers that cannot be accomplished with ciphers that only a recipient who knows the encoding method can decode.
4. Some encryption methods, such as Hill's Cipher, may be broken if small sections of plaintext and the corresponding ciphertext are known. Describe how someone attempting to decode a message might gain such knowledge about the encoded message.

PROJECTS

1. The Euclidean algorithm, described in the text, is a method for finding the greatest common divisor of two numbers.
   (a) Use the Euclidean algorithm to find gcd(1816046, 373678).
   (b) Write a spreadsheet or calculator program for computing gcd's using the Euclidean algorithm.
2. Develop your own check digit scheme for a serial number made up of nine digits, the last digit being a check digit. Determine which single-digit errors and errors consisting of the transposition of adjacent digits your scheme detects.
3. For each of the following check digit methods, determine which transposition errors go undetected:
   (a) banking routing number check digit method,
   (b) UPC check digit method.
4. Write a spreadsheet or computer program for carrying out the round-robin tournament method described in the text.
5. In some sports such as track, swimming, and golf, there are tri-meets where three teams face each other at the same time.
   (a) With 15 teams, it is possible to schedule 7 rounds of tri-meets so that each team plays in every round and faces each of the other 14 teams exactly once. Create such a schedule.
   (b) Find some basic conditions on the number of teams that must be satisfied in order to have any chance of creating such a schedule.
6. Letter frequency is useful in decoding. Two-letter block frequencies also prove useful, although there are $26 \times 26 = 676$ possible two-letter blocks. Find the two-letter block frequencies on a page of English text out of a book. What patterns appear frequently? Is this what you would expect? How could this information be used to decode a message encoded with the Hill cipher for two-letter blocks?

7. As you would expect, many companies that provide encryption software advertise on the Internet and elsewhere. These advertisements usually describe the system they use, the level of security they provide, the hardware they require, and the cost of the system. Investigate and report on some of these encryption systems.

8. The following passage was encoded with an affine cipher.

GMLAB   VHLMQ   MHZYH   LQHAN   MQMBD   MPCVM   JHHLW
HWBBS   MJWZM   EZMWH   MVMIY   WBHLW   HHLMO   WZMMJ
VAGMV   NOHLM   CZEZM   WHAZG   CHLEM   ZHWCJ   YJWBC
MJWNB   MZCUL   HQHLW   HWSAJ   UHLMQ   MWZMB   CDMBC
NMZHO   WJVHL   MRYZQ   YCHAD   LWRRC   JMQQH   LASWQ
TMDDM   ZQAJ

(a) Use the frequency distribution of letters in English given in Table 5.11 to decipher the passage.

(b) Find the key to the cipher.

9. Suppose you have intercepted the following message:

$$K\ X\ Z\ H\ I\ A\ M\ H\ G\ R\ K\ L\ B\ K\ G\ K\ C\ K\ F\ I\ O\ G\ W\ P.$$

Your spy network has discovered that the message was encoded by the Hill cipher with two-letter blocks and that the sender usually begins messages with "Dear." Find the key for decoding the cipher and decode the entire message.

10. In the Hill cipher for three-letter blocks, each block of three plaintext letters $P_1 P_2 P_3$ is encoded as a block of three ciphertext letters $C_1 C_2 C_3$, which are given by the three equations

$$C_1 = (aP_1 + bP_2 + cP_3) \bmod 26,$$

$$C_2 = (dP_1 + eP_2 + fP_3) \bmod 26,$$

$$C_3 = (gP_1 + hP_2 + iP_3) \bmod 26,$$

where a through i are integers between 0 and 25 that satisfy a condition that will make decoding possible. Find the decoding formula for this cipher and determine the necessary conditions on the integers a through i.

KEY TERMS

CHAPTER 5 / REVIEW TEST

1. Find the prime factorization of 4312.

2. Determine whether or not the given number is prime.
   (a) 437

   (b) 1553

3. Find the quotient and remainder when the division is performed according to the division algorithm.
   (a) 95 divided by 14

   (b) −115 divided by 7

4. Evaluate the given greatest common divisor.
   (a) gcd(90,126)

   (b) gcd(−46,920)

In Exercises 5–8, perform the given calculation. In each case, the answer should be a number between 0 and $n - 1$, where n is the modulus.

5. $(7 + 15 + 20) \bmod 9$

6. $(110 - 57) \bmod 25$

7. $(15^{300} + 13^{299}) \bmod 7$

8. $(753 \cdot 124 \cdot 43{,}921 \cdot 100{,}003) \bmod 10$

9. Use casting out nines to determine whether or not 42,098,793,390,115,866,045 is divisible by 9.

10. Use casting out elevens to determine whether or not 9,570,018,702,007 is divisible by 11.

11. Use the test for divisibility by 4 to determine whether or not 47,877,905,008,564,228 is divisible by 4.

12. Use the divisibility tests to determine whether or not the number 153,192,798,948 is divisible by each of the following numbers.

   (a) 2

   (b) 3

   (c) 4

   (d) 5

   (e) 6

   (f) 7

   (g) 8

   (h) 9

    (i)   10

    (j)   11

13. Eleven identical bags of potato chips were purchased by check. The checkbook entry reads \$20.#9, where the # is an illegible digit. Assuming there was no sales tax, how much did each bag of chips cost?

14. Consider the following calculation:

$$8{,}109{,}516{,}349 \cdot 4{,}510{,}902{,}130 = 36{,}581{,}234{,}571{,}973{,}023{,}370.$$

    (a) Use casting out nines to determine whether the calculation *could* be correct.

    (b) Use casting out elevens to determine whether the calculation *could* be correct.

15. The routing number of the Guaranty Federal Bank is 314970664. Suppose it is accidentally typed in as 319470664. Will the error be detected by the check digit?

16. Suppose the serial number for a U.S. Postal Service money order is 40983276587. If the number is accidentally recorded as 40983296587, will the check digit detect the error?

17. The main part of the 10-digit ISBN number of *Harry Potter and the Sorcerer's Stone* by J. K. Rowling is 0-590-35340. Find the check digit.

18. Suppose you know that the UPC number of a certain product is 07#628071504, where # is a digit that is illegible. What must the digit # be?

19. For a tournament with 22 teams, use our round-robin scheduling method to find the team that will be assigned to play Team 3 in round 18.

20. For a tournament with 19 teams, use our round-robin scheduling method to find the team that will be assigned to play Team 11 in round 6.

21. Decipher the following message, which was enciphered with the Caesar cipher:

    VKHNQ RZV.

22. Encode the message

    BRING SOME MONEY

    by breaking it into five-letter blocks and using the affine cipher $C = (7P + 20)$ mod 26.

23. Decipher the message

    BYPBE EBQAS JW,

    which was encoded using the affine cipher $C = (19P + 3)$ mod 26.

24. Encode the message

TURN LEFT

using the following Hill cipher for two-letter blocks:

$$C_1 = (9P_1 + 4P_2) \bmod 26,$$
$$C_2 = (5P_1 + 23P_2) \bmod 26.$$

25. Decipher the message

OR GJ NW TQ,

which was encoded using the following Hill cipher for two-letter blocks with some dummy letter in the last block:

$$C_1 = (8P_1 + 13P_2) \bmod 26,$$
$$C_2 = (3P_1 + 2P_2) \bmod 26.$$

26. Encode the message

BEGIN

using three-digit blocks and the RSA public key system where the message is being sent to the person with the public key address (3, 4189). Use the digit 0 for the dummy digits in the last block.

27. Decipher the message

2804 2170 0394 1206,

which was encoded using the RSA public key system with three-digit blocks and sent to the address (689, 3569), where $3569 = 43 \cdot 83$ and

$$\overline{689} \bmod ((43-1)(83-1)) = \overline{689} \bmod 3444 = 5.$$

SUGGESTED READINGS

Burton, David. *Elementary Number Theory.* New York: McGraw-Hill, 2010. An introductory number theory textbook.

Clausing, Jeri. "White House Yields a Bit on Encryption." *New York Times*, 8 July 1998, C1, 5. A brief report on the national security and business sides of the debate over the export of cryptographic technology.

Crandall, Richard E. "The Challenge of Large Numbers." *Scientific American* 276:2 (February 1997), 74–78. Large numbers in history and how computers calculate with large numbers, emphasizing number theory and cryptology.

Dewdney, A. K. "On Making and Breaking Codes: Part I." *Scientific American* 259:4 (October 1988), 144–147; "On Making and Breaking Codes: Part II." *Scientific American* 259:5 (November 1988), 142–145. Letter frequencies and the German Enigma machine in Part I, the Data Encryption Standard and the RSA cryptosystem in Part II.

Gallian, Joseph. "Assigning Driver's License Numbers." *Mathematics Magazine* 64:1 (February 1991), 13–22. Methods used by states to assign driver's license numbers, including several check digit schemes.

Gallian, Joseph. "The Mathematics of Identification Numbers." *College Mathematics Journal* 22:3 (May 1991), 194–202. A more technical look at some check digit schemes. Includes a description of Verhoeff's check digit method.

Gallian, Joseph, and Steven Winters. "Modular Arithmetic in the Marketplace." *American Mathematical Monthly* 95:6 (June–July 1988), 548–551. A more technical look at some check digit schemes.

Garrett, Paul. *Making, Breaking Codes: Introduction to Cryptology.* Upper Saddle River, New Jersey: Prentice Hall, 2001. An introduction to classical and modern cryptology, including the RSA public key system.

Singh, Simon. *The Code Book: The Science of Secrecy from Ancient Egypt to Quantum Cryptography.* New York: Anchor Books, 2000. An entertaining history of encryption including its effect on wars, politics, and internet security.

Singh, Simon, and Kenneth A. Ribet. "Fermat's Last Stand." *Scientific American* 277:5 (November 1997), 68–73. The history of Fermat's Last Theorem and the story of Andrew Wiles solving the problem.

ANSWERS TO MOST ODD-NUMBERED EXERCISES AND CHAPTER REVIEW TESTS

CHAPTER 1

Section 1.1
1. Chinese **3. (a)** 12 **(b)** 20 **(c)** 23 **5.** 62 **7. (a)** Jackson or Adams **(b)** 14.62% **9.** 36.71%
11. (a) carpet **(b)** wood **13. (a)** no curfew **(b)** 1 A.M. **15. (a)** 40 hours **(b)** 40 hours **(c)** no **(d)** Yes, they could vote so that 20 hours wins. **17. (a)** volleyball **(b)** Yes, they could vote so that softball wins.
19. volleyball **21.** softball **23.** yes **25.** football

Section 1.2
1. Cardona **3.** Horned Frogs **5.** Monday **7.** Tilson **9. (a)** Grunick **(b)** tie between Burgundy and Madden **(c)** tie between Burgundy and Madden **11. (a)** football **(b)** football **(c)** football **(d)** Yes, they could vote so that baseball wins. **13. (a)** Cullors **(b)** Yes, they could vote so that Allen wins. **(c)** no

Section 1.3
1. (a) Sanders **(b)** Bauer **(c)** Sanders **(d)** Sanders **3.** Mexican **5.** none **7.** Hungry Boar
9. (a) optional uniforms **(b)** no uniforms **(c)** no uniforms **(d)** optional uniforms **11. (a)** beagle
(b) beagle **(c)** beagle **(d)** none **13.** no **15. (a)** public development **(b)** private development
(c) recommitment **(d)** none **17. (a)** and **(b)** **19.** yes

Section 1.4
1. Freeman **3. (a)** A **(b)** 0.956% **5.** De Castro **7. (a)** eighteen **(b)** eighteen **(c)** eighteen **(d)** eighteen
(e) eighteen **9. (a)** Bailey **(b)** Holmes **(c)** Holmes **(d)** Holmes **(e)** Bailey **(f)** no **11. (a)** white **(b)** ivory
(c) tie between white and ivory **(d)** ivory **(e)** ivory **(f)** Yes, they could vote so that white wins.
13. Lieberman **15.** Lieberman **17. (b)** 9.4% **(c)** 15.3% **(d)** 42.3% **(e)** Reagan 14%, Carter 23%, Anderson 63% **(f)** Reagan 217.2, Carter 206.4, Anderson 176.4; Reagan wins **19. (a)** Wilson **(b)** Wilson
(c) Wilson **(d)** Roosevelt **(e)** Wilson **(f)** Wilson **21. (a)** Taft **(b)** Roosevelt **(c)** Wilson **(d)** Roosevelt
(e) Wilson **(f)** Roosevelt **23. (a)** Baldwin **(b)** Baldwin **(c)** Baldwin **(d)** Baldwin **25. (a)** Musselman
(b) Watkins **(c)** Watkins **(d)** Watkins **27. (a)** Buckley **(b)** Goodell **(c)** Goodell

Chapter 1 Review Test
1. 77 **2. (a)** 18 **(b)** 27 **3. (a)** bananas **(b)** bananas **(c)** grapes **(d)** bananas **4. (a)** basketball court
(b) baseball field **(c)** baseball field **(d)** baseball field **5.** jazz trio **6. (a)** Romeo and Juliet **(b)** Romeo and Juliet **(c)** The Fantasticks **(d)** none **(e)** no **(f)** Yes, they could vote so that Death of a Salesman wins.
7. (a) tie between Mintz and Zukoff **(b)** Zukoff **(c)** Zukoff **(d)** Zukoff **(e)** Mintz **(f)** Yes, they could vote so that Mintz wins. **(g)** Yes, they could vote so that Zukoff wins. **8.** If every voter prefers A to B, then B has no first-place votes and cannot make the runoff, so cannot win the election.
9. One example is

Number of Voters			
	1	1	1	1
A	1	4	3	2
B	2	1	4	3
C	3	2	1	4
D	4	3	2	1

CHAPTER 2

Section 2.1
1. 2.02319948 **3.** 1.04035789 **5.** 1.84639651 **7.** 6.06388151 **9.** −3.79376085 **11.** −0.31546488 **13.** 0.93578497 **15.** 0.17216130 **17.** 0.01857668 **19.** 0.00921670 **21.** 12.66977089 **23.** 0.01729323 **25.** 3.02246118 **27.** 19.59094589 **29.** 10.75121514

Section 2.2
1. $600 **3.** $727.94 **5.** $410.67 **7.** $358.40 **9.** $51,530.24 **11.** $1913.88 **13.** $57,397.96 **15.** 5.38 years **17.** 16.59% **19.** $2237.93 **21.** 456.41% **23.** 1.82 years **25.** 4320 talents **27.** $137.50 **29.** $560 **31.** 485.45%

Section 2.3
1. $7895.59 **3.** $9750.44 **5.** $2458.65 **7.** $1151.61 **9.** $4197.54 **11.** $150.56 **13.** 8.83 years **15.** 15.58 years **17.** 10.67% **19.** 6.58% **21.** 4.80% **23.** 4.03% **25.** (a) 5.09% (b) 5.12% (c) 5.13% **27.** (a) $793.44 (b) $802.41 (c) $804.52 (d) $805.55 **29.** Ascencia Bank **31.** 11.31% **33.** $0.93835141 million or $938,351.41 **35.** $378.05 **37.** 3.59 years **39.** $21,631.96 **41.** $71,135.00 **43.** $83,856.13 **45.** 17.68 years **47.** $1755.44 **49.** (a) $643,000 (b) $2.62172775 · 10^{31} **51.** $17,720.09 **53.** (a) $52,214.57 (b) 1.36%

Section 2.4
1. $59,561.68 **3.** $151,898.73 **5.** $687.42 **7.** $222.61 **9.** 1.29 years **11.** 5.53 years **13.** (a) $14,513.57 (b) $9000.00 (c) $5513.57 **15.** 2.62 years **17.** (a) $471,591.05 (b) $243,994.01 **19.** (a) $466.11 (b) $1787.55 **21.** lump sum option: $80,979.25; $420 payments for 5 years option: $80,089.71; $182 payments for 17 years, 4 months option: $79,751.69 **23.** $48,478.85 **25.** (b) $1725.92

Section 2.5
1. $14,332.92 **3.** $1731.54 **5.** $862.00 **7.** $1383.07 **9.** 3.15 years **11.** 8.42 years **13.** $7923.19 **15.** (a) $95.46 (b) $11,455.20 (c) $2455.20 **17.** $2330.74 **19.** 3.03 years **21.** (a) $701.77 (b) $75,746.82 (c) $52,815.61 **23.** the 0.9% option **25.** (a) $1055.47 (b) 17.69 years (c) $113,458.03 **27.** (a) $472.81 (b) $700.24 (c) $44,168.40

29.

Payment Number	Payment	Interest Paid	Principal Paid	Balance
				8000.00
1	267.42	120.00	147.42	7852.58
2	267.42	117.79	149.63	7702.95
3	267.42	115.54	151.88	7551.07
4	267.42	113.27	154.15	7396.92

31.

Payment Number	Payment	Interest Paid	Principal Paid	Balance
				5429.38
1	927.20	38.01	889.19	4540.19
2	927.20	31.78	895.42	3644.77
3	927.20	25.51	901.69	2743.08
4	927.20	19.20	908.00	1835.08
5	927.20	12.85	914.35	920.73
6	927.18	6.45	920.73	0

33.

Payment Number	Payment	Interest Paid	Principal Paid	Balance
				7300.00
1	2574.57	209.88	2364.69	4935.31
2	2574.57	141.89	2432.68	2502.63
3	2574.58	71.95	2502.63	0

35. (a) $346,604.40 **(b)** $103,651.91 **(c)** $331,419.60 **(d)** yes **37. (a)** $14,594.40 **(b)** yes **(c)** $972.96 **(d)** $187,538.42 **(e)** $187,981.04 **(f)** no **39. (a)** yes **(b)** no **41.** $54,772.73 **43.** 6.67 years **45.** $439,474.67 **47. (a)** $222,245.46 **(b)** $6,667,363.80 **49. (b)** $368.64

Chapter 2 Review Test

1. $907.33 **2.** 80% **3. (a)** $666,876.87 **(b)** $92,000.00 **(c)** $574,876.87 **4. (a)** $3500.29 **(b)** $294,024.36 **(c)** $70,877.36 **5.** $3434.64 **6.** 17.35 years **7. (a)** $194,894.75 **(b)** $244,222.90 **8.** $815.75 **9.** 27,000 pounds

10.

Payment Number	Payment	Interest Paid	Principal Paid	Balance
				943.61
1	160.87	6.13	154.74	788.87
2	160.87	5.13	155.74	633.13
3	160.87	4.12	156.75	476.38
4	160.87	3.10	157.77	318.61
5	160.87	2.07	158.80	159.81
6	160.85	1.04	159.81	0

11. (a) $15,485.54 **(b)** $4461.28   **12.** $30,120.22   **13. (a)** $172.83 **(b)** 6.92 years **(c)** $2235.80
14. 18.73%   **15.** 1.156%   **16.** 1.30 years   **17.** 5.46 years   **18.** 10% down payment option:
$31,430.08; 20% down payment option: $30,556.16   **19.** $4674.94   **20. (a)** $185,159.56 **(b)** 6.60%
(c) $199,334.33 **(d)** the lump sum payment option

CHAPTER 3

Section 3.1
1. A, B **3.** B, C, D, E **5.** It has an Eulerian circuit. One possibility is as follows:

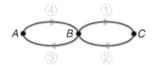

7. It has four odd vertices, A, B, C, and D, so it has neither an Eulerian circuit nor an Eulerian path.
9. It has an Eulerian circuit. One possibility is as follows:

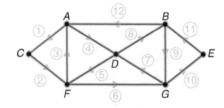

11. It has two odd vertices, A and D, so it does not have an Eulerian circuit. It does have an Eulerian
path. One possibility is as follows:

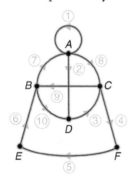

13. It has two odd vertices, A and C, so it does not have an Eulerian circuit. It does have an Eulerian
path. One possibility is as follows:

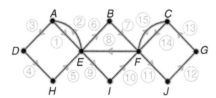

15. It is possible to take such a walk. One such walk is as follows:

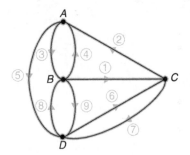

17. The graph has an Eulerian circuit. One possibility is as follows:

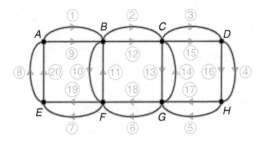

19. The graph has two odd vertices, *B* and *F*, so it does not have an Eulerian circuit. It does have an Eulerian path. One possibility is as follows:

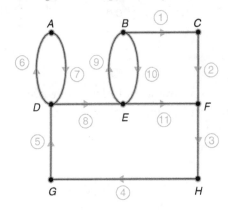

21. (a)

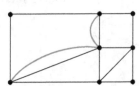

(b) The graph has two odd vertices, *A* and *D*, so it is not possible.

23. One possible eulerization is as follows:

25. One possible eulerization is as follows:

27. One possible eulerization is as follows:

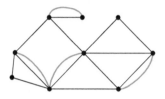

29. One possible eulerization is as follows:

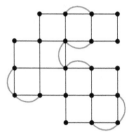

31. (a)

(b) One possible eulerization is as follows:

(c) The eulerization shown in (b) gives the shortest tour of the park.

Section 3.2

1. One Hamiltonian circuit is shown here.

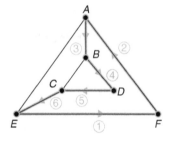

3. One Hamiltonian circuit is shown here.

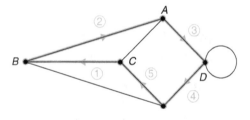

5. There are no Hamiltonian circuits on this graph because any circuit would have to go through the edge between vertices D and E twice and therefore visit both vertex D and vertex E twice.

7. (a)

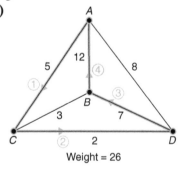

(b)

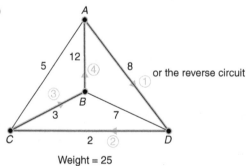

9. (a)

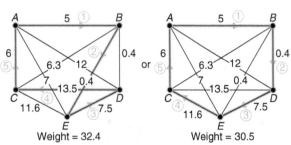

(b)

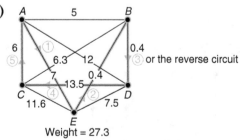

11. (a)

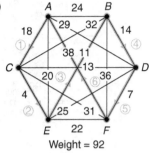

Weight = 92

(b)

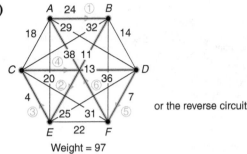

or the reverse circuit

Weight = 97

13. (a)

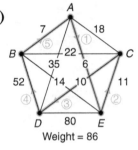

Weight = 86

(b) the same as the circuit in part (a) or the reverse circuit; weight = 86

15. (a)

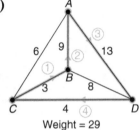

Weight = 30

(b)

Weight = 29

(c)

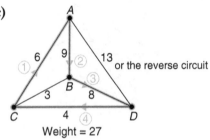

or the reverse circuit

Weight = 27

17. (a)

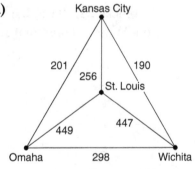

(b)

Length = 1287 miles

(c) the same as the circuit in part (b) or the reverse circuit; length = 1287 miles

(d) It is 94 miles shorter.

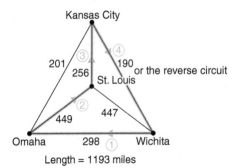

or the reverse circuit

Length = 1193 miles

19. (a)

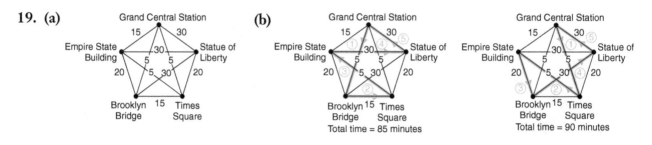

(b)

(c)

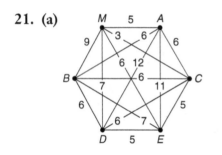

or the reverse circuit

21. (a)

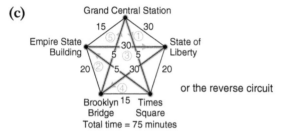

(b)

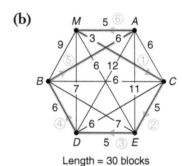

(c) the same as the circuit in part (a) or the reverse circuit; length = 30 blocks **23.** Base, Barnard 3, Shoemaker–Levy 7, Gunn, Swift, Neujmin 2, du Toit–Hartley, Harrington–Wilson, Base; total angular change = 419.9°

25. (a)

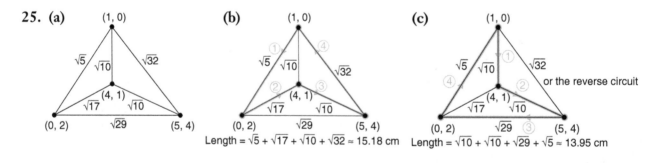

27. One Hamiltonian circuit is as follows:

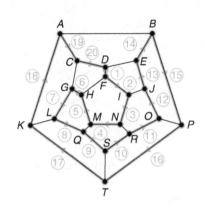

29.

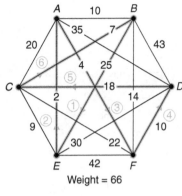

Weight = 66

Section 3.3

1. not a tree because it includes a circuit   **3.** a tree, but not a spanning tree   **5.** a spanning tree

7. Three possibilities are as follows:

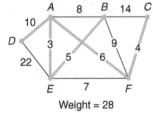

9.

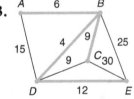

11.

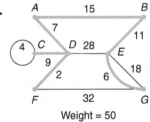

Weight = 28

13.

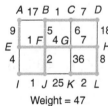

or

Weight = 31

15.

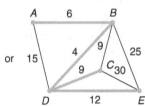

Weight = 50

17.

A 17 B 1 C 7 D
9   5   6   18
E 1 F 4 G 7 H
4   2   36   8
I 1 J 25 K 2 L

Weight = 47

19.

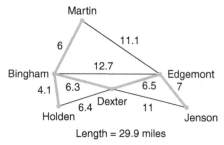

Length = 29.9 miles

21. The following flights should be retained.

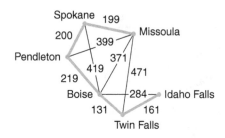

23. The cable should be installed as shown here.

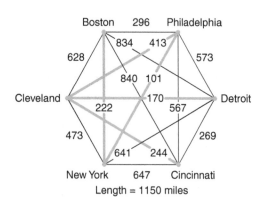

Length = 1150 miles

25. The minimum cost network is as follows:

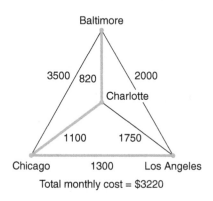

Total monthly cost = $3220

Chapter 3 Review Test

1. It has two odd vertices, A and B, so it does not have an Eulerian circuit. It has an Eulerian path. One possibility is as follows:

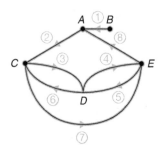

2. It has four odd vertices, B, C, D, and I, so it has neither an Eulerian circuit nor an Eulerian path.

3. It has six odd vertices, A, B, C, E, F, and H, so it has neither an Eulerian circuit nor an Eulerian path.

4. It has an Eulerian circuit. One possibility is as follows:

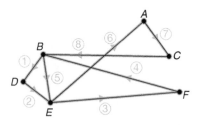

5. One possible eulerization is shown here.

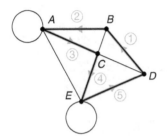

6. One possible eulerization is shown here.

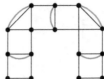

7. One Hamiltonian circuit is shown here.

8. It does not have a Hamiltonian circuit because any circuit must go through vertex *C* at least twice.

9. (a)

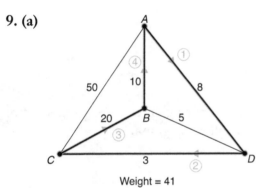

Weight = 41

(b)

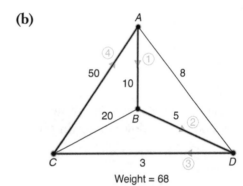

Weight = 68

10. (a)

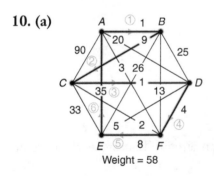

Weight = 58

(b)

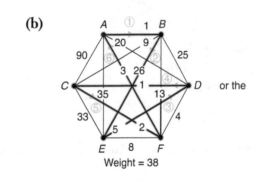

or the

Weight = 38

11.

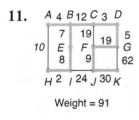

Weight = 91

12.

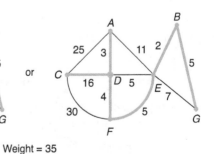

or

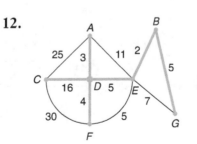

Weight = 35

13. The graph has an Eulerian circuit. One possibility is as follows:

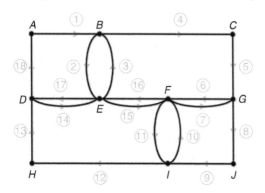

14. We first label the rooms of the house and the area outside the house as shown here.

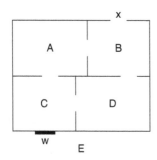

Letting the rooms and the area outside the house be vertices and drawing an edge between any two vertices that have a door or window connecting them, we get the following graph.

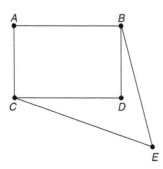

This graph has two odd vertices, *B* and *C*, so it cannot have an Eulerian circuit. Therefore, the burglar must be lying.

15. (a) **(b)** **(c)**

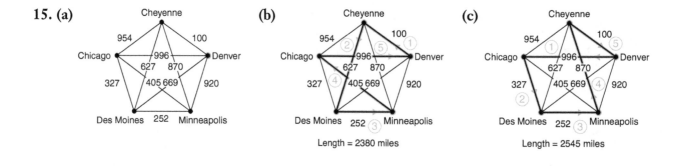

16. The cheapest way of installing the pipes is shown here.   **17**. for *n* odd

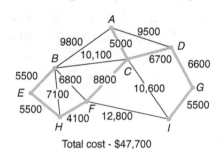

Total cost - $47,700

CHAPTER 4

Section 4.1

1. yes   **3**. no   **5**. no   **7**. no   **9**. concave; not regular; heptagon   **11**. convex; regular; nonagon
13. convex; not regular; quadrilateral   **15**. convex; regular; 11-gon   **17**. 1440°   **19**. 2700°
21. 1260°   **23**. 135°   **25**. 128.5714°   **27**. 168°   **29**. 21°   **31**. 185°

Section 4.2

1. not edge-to-edge   **3**. Not all vertices are of the same type.   **5**. The tiles are not all regular polygons
and not all vertices are of the same type.   **7**. The four different vertex types are shown here.

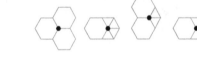

9. The two different vertex types are shown here.

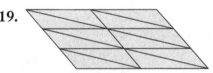

17. an equilateral triangle and a regular octagon   **19**.

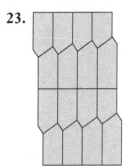

21.

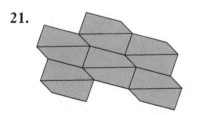

23.

Section 4.3

1. convex   **3.** concave   **5.** $V - E + F = 10 - 17 + 9 = 2$ **7.** $V - E + F = 8 - 14 + 8 = 2$
9. $V - E + F = 4 - 6 + 4 = 2$ **11.** $V - E + F = 6 - 12 + 8 = 2$ **13.** $V - E + F = 12 - 30 + 20 = 2$
15. 12   **17.** All of the faces of the polyhedron are triangles.

19. The polyhedron either has 4 triangular faces and 2 quadrilateral faces or it has 5 triangular faces and 1 pentagonal face.

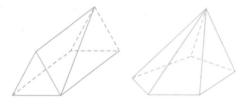

21. **(a)** 14 **(b)** 8 **(c)**

23. $V - E + F = 16 - 24 + 10 = 2$ **25.** $V = 16, E = 32, F = 16$ **27.** $V = 40, E = 84,$
$F = 40$ **31. (a)** 8 **(b)** 6 **33.** 4 equilateral triangle faces and 4 regular hexagon
faces   **35.** cube   **37.** dodecahedron   **43.** $V = 6, E = 12; V = 7, E = 13; V = 8, E = 14;$
$V = 9, E = 15; V = 10, E = 16; V = 11, E = 17; V = 12, E = 18$

Chapter 4 Review Test

1.    **2.** 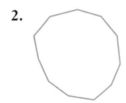   **3.** 156°   **4.** 86°

5.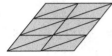

6.

7. The three different vertex types are shown here. **8.** 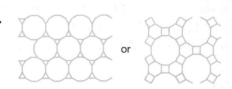 or

9. an equilateral triangle and a regular heptagon **10.** The other polygons about vertex A must be two hexagons. However, this forces hexagons and triangles to alternate about vertex B, a different vertex type.

11. One of the vertices is surrounded by a pentagon, an octagon, and a hexagon. If they were all regular, the sum of the angles would be

$$\left(180° - \frac{360°}{5}\right) + \left(180° - \frac{360°}{8}\right) + \left(180° - \frac{360°}{6}\right) = 108° + 135° + 120° = 363°.$$

Because the angles do not sum to 360°, these polygons cannot all be regular.

12. $V - E + F = 7 - 12 + 7 = 2$ **13.** 25 **14.** 24 **15.** 1 quadrilateral face and 4 triangular faces

16. (a) 20 (b) 12 (c) $V - E + F = 60 - 90 + 32 = 2$

CHAPTER 5

Section 5.1

1. T **3.** F **5.** F **7.** F **9.** T **11.** T **13.** $2^2 \cdot 3 \cdot 11$ **15.** $5^3 \cdot 13$ **17.** $2^4 \cdot 5^2 \cdot 17$ **19.** 2, 3, 5, 7, 11, 13, 17, 19, 23, 29, 31, 37, 41, 43, 47, 53, 59, 61, 67, 71, 73, 79, 83, 89, 97 **21.** prime **23.** not prime **25.** prime **29.** 3 + 13 or 5 + 11 **31.** 3 + 43 or 5 + 41 or 17 + 29 or 23 + 23 **33.** 19 + 79 or 31 + 67 or 37 + 61 **35.** One possibility is 86 = 7 + 13 + 23 + 43. **37.** One possibility is 232 = 3 + 11 + 19 + 31 + 37 + 41 + 43 + 47. **39.** $q = 9, r = 5$ **41.** $q = 17, r = 0$ **43.** $q = -8, r = 4$ **45.** $q = -14,$ $r = 3$ **47.** $q = -9, r = 3$ **49.** 3 **51.** 24 **53.** 8 **55.** 1 **57.** 25 **59.** 21 **65.** (a) all positive integers (b) integers of the form $4k$ or $4k-1$, where k is a positive integer

Section 5.2

1. 3 **3.** 10 **5.** 4 **7.** 0 **9.** 0 **11.** 2 **13.** 7 **15.** 1 **17.** 7 **19.** 19 **21.** 3 **23.** 8 **25.** 2 **27.** 3 **29.** 6 **31.** 2 **33.** 5 **35.** 18 **37.** 1 **39.** 43 **41.** 9 **43.** 1 **45.** 9803

Section 5.3

1. divisible by 9 **3.** not divisible by 9 **5.** divisible by 9 **7.** not divisible by 11 **9.** not divisible by 11 **11.** divisible by 11 **13.** not divisible by 4 **15.** divisible by 4 **17. (a)** yes **(b)** yes **(c)** yes **(d)** no **(e)** yes **(f)** no **(g)** yes **(h)** no **(i)** no **(j)** no **19. (a)** no **(b)** yes **(c)** no **d)** yes **(e)** no **(f)** yes **(g)** no **(h)** yes **(i)** no **(j)** no **25.** 49¢ **29.** It could be correct. **31.** It could not be correct.

Section 5.4

1. 64867456697 **3.** no **5.** yes **7.** 1 **9.** 6 **11.** no **13.** 4 **15.** 6 **17.** yes **19. (a)** no **(b)** no **21.** K **23.** yes

Section 5.5

1.

	\<span\>Round\</span\>								
	1st	2nd	3rd	4th	5th	6th	7th	8th	9th
Team 1	9	Bye	2	3	4	5	6	7	8
Team 2	8	9	1	Bye	3	4	5	6	7
Team 3	7	8	9	1	2	Bye	4	5	6
Team 4	6	7	8	9	1	2	3	Bye	5
Team 5	Bye	6	7	8	9	1	2	3	4
Team 6	4	5	Bye	7	8	9	1	2	3
Team 7	3	4	5	6	Bye	8	9	1	2
Team 8	2	3	4	5	6	7	Bye	9	1
Team 9	1	2	3	4	5	6	7	8	Bye

3.

	Round								
	1st	2nd	3rd	4th	5th	6th	7th	8th	9th
Team 1	9	10	2	3	4	5	6	7	8
Team 2	8	9	1	10	3	4	5	6	7
Team 3	7	8	9	1	2	10	4	5	6
Team 4	6	7	8	9	1	2	3	10	5
Team 5	10	6	7	8	9	1	2	3	4
Team 6	4	5	10	7	8	9	1	2	3
Team 7	3	4	5	6	10	8	9	1	2
Team 8	2	3	4	5	6	7	10	9	1
Team 9	1	2	3	4	5	6	7	8	10
Team 10	5	1	6	2	7	3	8	4	9

5.

							Round								
	1st	2nd	3rd	4th	5th	6th	7th	8th	9th	10th	11th	12th	13th	14th	15th
Team 1	15	Bye	2	3	4	5	6	7	8	9	10	11	12	13	14
Team 2	14	15	1	Bye	3	4	5	6	7	8	9	10	11	12	13
Team 3	13	14	15	1	2	Bye	4	5	6	7	8	9	10	11	12
Team 4	12	13	14	15	1	2	3	Bye	5	6	7	8	9	10	11
Team 5	11	12	13	14	15	1	2	3	4	Bye	6	7	8	9	10
Team 6	10	11	12	13	14	15	1	2	3	4	5	Bye	7	8	9
Team 7	9	10	11	12	13	14	15	1	2	3	4	5	6	Bye	8
Team 8	Bye	9	10	11	12	13	14	15	1	2	3	4	5	6	7
Team 9	7	8	Bye	10	11	12	13	14	15	1	2	3	4	5	6
Team 10	6	7	8	9	Bye	11	12	13	14	15	1	2	3	4	5
Team 11	5	6	7	8	9	10	Bye	12	13	14	15	1	2	3	4
Team 12	4	5	6	7	8	9	10	11	Bye	13	14	15	1	2	3
Team 13	3	4	5	6	7	8	9	10	11	12	Bye	14	15	1	2
Team 14	2	3	4	5	6	7	8	9	10	11	12	13	Bye	15	1
Team 15	1	2	3	4	5	6	7	8	9	10	11	12	13	14	Bye

7.

							Round								
	1st	2nd	3rd	4th	5th	6th	7th	8th	9th	10th	11th	12th	13th	14th	15th
Team 1	15	16	2	3	4	5	6	7	8	9	10	11	12	13	14
Team 2	14	15	1	16	3	4	5	6	7	8	9	10	11	12	13
Team 3	13	14	15	1	2	16	4	5	6	7	8	9	10	11	12
Team 4	12	13	14	15	1	2	3	16	5	6	7	8	9	10	11
Team 5	11	12	13	14	15	1	2	3	4	16	6	7	8	9	10
Team 6	10	11	12	13	14	15	1	2	3	4	5	16	7	8	9
Team 7	9	10	11	12	13	14	15	1	2	3	4	5	6	16	8
Team 8	16	9	10	11	12	13	14	15	1	2	3	4	5	6	7
Team 9	7	8	16	10	11	12	13	14	15	1	2	3	4	5	6
Team 10	6	7	8	9	16	11	12	13	14	15	1	2	3	4	5
Team 11	5	6	7	8	9	10	16	12	13	14	15	1	2	3	4
Team 12	4	5	6	7	8	9	10	11	16	13	14	15	1	2	3
Team 13	3	4	5	6	7	8	9	10	11	12	16	14	15	1	2
Team 14	2	3	4	5	6	7	8	9	10	11	12	13	16	15	1
Team 15	1	2	3	4	5	6	7	8	9	10	11	12	13	14	16
Team 16	8	1	9	2	10	3	11	4	12	5	13	6	14	7	15

9. Team 17 **11.** Team 21 **13.** Round 17 **15.** Round 7

17. The schedule is given in Exercise 3. Home teams: Round 1: 1, 2, 3, 4, 10; Round 2: 6, 7, 8, 9, 10; Round 3: 2, 3, 4, 5, 6; Round 4: 1, 2, 7, 8, 9; Round 5: 3, 4, 5, 6, 10; Round 6: 1, 2, 8, 9, 10; Round 7: 4, 5, 6, 7, 8; Round 8: 1, 2, 3, 4, 9; Round 9: 5, 6, 7, 8, 10

19. The schedule is given in Exercise 5. Home teams: Round 1: 1, 2, 3, 4, 5, 6, 7; Round 2: 9, 10, 11, 12, 13, 14, 15; Round 3: 2, 3, 4, 5, 6, 7, 8; Round 4: 1, 10, 11, 12, 13, 14, 15; Round 5: 3, 4, 5, 6, 7, 8, 9; Round 6: 1, 2, 11, 12, 13, 14, 15; Round 7: 4, 5, 6, 7, 8, 9, 10; Round 8: 1, 2, 3, 12, 13, 14, 15; Round 9: 5, 6, 7, 8, 9, 10, 11; Round 10: 1, 2, 3, 4, 13, 14, 15; Round 11: 6, 7, 8, 9, 10, 11, 12; Round 12: 1, 2, 3, 4, 5, 14, 15; Round 13: 7, 8, 9, 10, 11, 12, 13; Round 14: 1, 2, 3, 4, 5, 6, 15; Round 15: 8, 9, 10, 11, 12, 13, 14

23. The schedule is given in Exercise 3. Home teams: Round 1: 3, 4, 5, 8, 9; Round 2: 3, 4, 5, 9, 10; Round 3: 1, 4, 5, 6, 9; Round 4: 1, 4, 5, 6, 10; Round 5: 1, 2, 5, 6, 7; Round 6: 1, 2, 6, 7, 10; Round 7: 1, 2, 3, 7, 8; Round 8: 2, 3, 7, 8, 10; Round 9: 2, 3, 4, 8, 9

25. The schedule is given in Exercise 5. Home teams: Round 1: 5, 6, 7, 12, 13, 14, 15; Round 2: 5, 6, 7, 8, 13, 14, 15; Round 3: 1, 6, 7, 8, 13, 14, 15; Round 4: 1, 6, 7, 8, 9, 14, 15; Round 5: 1, 2, 7, 8, 9, 14, 15; Round 6: 1, 2, 7, 8, 9, 10, 15; Round 7: 1, 2, 3, 8, 9, 10, 15; Round 8: 1, 2, 3, 8, 9, 10, 11; Round 9: 1, 2, 3, 4, 9, 10, 11; Round 10: 2, 3, 4, 9, 10, 11, 12; Round 11: 2, 3, 4, 5, 10, 11, 12; Round 12: 3, 4, 5, 10, 11, 12, 13; Round 13: 3, 4, 5, 6, 11, 12, 13; Round 14: 4, 5, 6, 11, 12, 13, 14; Round 15: 4, 5, 6, 7, 12, 13, 14

29.

	Round						
	1st	**2nd**	**3rd**	**4th**	**5th**	**6th**	**7th**
Team 1	Bye	3	5	7	2	4	6
Team 2	7	Bye	4	6	1	3	5
Team 3	6	1	Bye	5	7	2	4
Team 4	5	7	2	Bye	6	1	3
Team 5	4	6	1	3	Bye	7	2
Team 6	3	5	7	2	4	Bye	1
Team 7	2	4	6	1	3	5	Bye

The rounds have been scrambled.

35.

	Round						
	1st	**2nd**	**3rd**	**4th**	**5th**	**6th**	**7th**
Team 1	2	3	4	5	6	7	8
Team 2	1	8	3	4	5	6	7
Team 3	7	1	2	8	4	5	6
Team 4	6	7	1	2	3	8	5
Team 5	8	6	7	1	2	3	4
Team 6	4	5	8	7	1	2	3
Team 7	3	4	5	6	8	1	2
Team 8	5	2	6	3	7	4	1

Section 5.6

1. DQBWL PHLVR NDB   **3.** YOU GOT IT RIGHT   **5.** EGKAX GKRDA FOE   **7.** KFSFD LEKFL FTSFE   **9.** PLAY IT AGAIN   **11.** DON'T THINK TWICE

Section 5.7

1. CM OR AD IC YW UM SR   **3.** SK NE PY OX   **5.** VOTE TODAY   **7.** HIT THE ROAD **9.** 1457 0434 1036 1420 0843 1528   **11.** 3521 4185 3906 4101 0867 2704   **13.** PARDON ME   **15.** BE SURE

Chapter 5 Review Test

1. $2^3 \cdot 7^2 \cdot 11$ **2.** **(a)** not prime **(b)** prime   **3.** **(a)** $q = 6, r = 11$ **(b)** $q = -17, r = 4$ **4.** **(a)** 18 **(b)** 46   **5.** 6 **6.** 3   **7.** 0   **8.** 6   **9.** divisible by 9   **10.** not divisible by 11   **11.** divisible by 4   **12.** **(a)** yes **(b)** yes **(c)** yes **(d)** no **(e)** yes **(f)** yes **(g)** no **(h)** no **(i)** no **(j)** no   **13.** $1.89   **14.** **(a)** It could be correct. **(b)** It could not be correct.   **15.** no   **16.** yes   **17.** 3   **18.** 8   **19.** Team 15   **20.** Team 14 21. SHE KNOWS   **22.** BJYHK QOAWA OHWG   **23.** EXCELLENT JOB   **24.** RJ XU LR RU   **25.** FOR SALE   **26.** 1000 4141 3627 1895   **27.** BE NICE